ENERGY, ECONOMICS, AND POLITICS IN THE CASPIAN REGION

ENERGY, ECONOMICS, AND POLITICS IN THE CASPIAN REGION

DREAMS AND REALITIES

Maureen S. Crandall

PRAEGER SECURITY INTERNATIONAL
Westport, Connecticut • London

Library of Congress Cataloging-in-Publication Data

Crandall, Maureen S.

Energy, economics, and politics in the Caspian region : dreams and realities / Maureen S. Crandall.

p. cm.

Includes bibliographical references and index.

ISBN 0–275–99130–X (alk. paper)

1. Petroleum industry and trade—Caspian Sea Region. 2. Gas industry—Caspian Sea Region. 3. Caspian Sea Region—Politics and government. 4. Caspian Sea Region—Economic conditions. 5. Geopolitics—Caspian Sea Region. I. Title.

HD9576.C372C73 2006

333.8′2309475–dc22 2006008228

British Library Cataloguing in Publication Data is available.

Library of Congress Catalog Card Number: 2006008228
ISBN: 0–275–99130–X

First published in 2006

Praeger Security International, 88 Post Road West, Westport, CT 06881
An imprint of Greenwood Publishing Group, Inc.
www.praeger.com

Printed in the United States of America

The paper used in this book complies with the Permanent Paper Standard issued by the National Information Standards Organization (Z39.48–1984).

10 9 8 7 6 5 4 3 2 1

Contents

Acknowledgments

This study owes a lot to many. My thanks go first to the National Defense University and the Industrial College of the Armed Forces (ICAF) for giving me the time to think and write about this subject. Among my most important supporters has been my Economics Department Chairwoman at ICAF, Professor Irene Kyriakopoulos, who was always there when I so greatly needed her support, regardless of the reason. She talked me through a lot of problems along the way. She was ably assisted by the then–Dean of Academic Affairs Dr. John Yaeger (Captain, U.S. Navy, retired). Similarly, my ICAF Economics Department colleague, Professor David King was, before and after the research began, a rock of help and encouragement, and always a willing listener, critic, and professional and personal conversationalist. Dave is in many ways my alter ego—he helps finish sentences and thoughts, is a wonderful personal and professional colleague, and is always a good friend. Without Dave's careful attention, the prolonged effort would have been nearly intolerable.

Beyond these, there is a special and long-term debt of thanks to Dr. Charles Collier, whom I have known both personally and professionally for more than 20 years. It was he who made the always-in-good-spirit comments about how the organization and exposition could be improved. His humor was remarkable as he conducted a tough job of review. But most of all, he is a walking encyclopedia on this part of the world, and served as a very important touchstone for both fact and interpretation. Anyone interested in this subject and geographic region needs to consult regularly with Dr. Collier.

Deep and sincere thanks go to the outstanding staff at the National Defense University library, all of whom were truly selfless in their willing responses to some of the most abstruse, maybe even obtuse, questions that they have ever received. Their total professionalism and outstanding demeanors are traits that mark them in all their interactions with faculty and students, and have given me memories I always will treasure. These wonderful persons include Julie Arighetti, Jeanne-Marie Faison, Elizabeth Harrison, Crafton Hayes, Howard Hume, Kim Jordan, Rosemary Marlowe-Dziuk, Lily McGovern, Sarah Mikel, Robert Montgomery, Bruce Thornlow, Jose Torres, and Cheryl Weidner. In particular, there is one at the library who adopted me, and I, her—she is Dawn Humphrey, without whom this work could never have been completed. She is a gem beyond compare. I wish her well in every respect, and those ICAF faculty who follow me on sabbatical leave should know that Dawn is truly a national treasure.

In addition, the work benefited greatly from the gentle and capable assistance of Michael Harn in ICAF's computer support group. His knowledge of computers and their capabilities is encyclopedic, and he was indispensable in making the tables and graphics useable and presentable. Moreover, his good cheer was endless. Would that I could emulate his IT skills.

In the course of this research, I emailed and spoke with a number of individuals. While I have tried to protect the identity of these most helpful people, there are at least three who outshine them all, in terms of their personal and professional assistance. These are Adam Sieminski of Deutsche Bank, Len Coburn, formerly with the U.S. Department of Energy, and Peter Davies of BP. Always a source of knowledge, support, and good cheer, Adam's encouragement and assistance was invaluable. Len has a depth of understanding and candor that is hard to match, and he shared it willingly. Peter stimulated interest in this area in the first place, together with BP's Michael Townshend. Adam, Peter, and Michael have their own views, but were willing to share and probe. Special thanks go to Ambassador William Ramsay at the International Energy Agency in Paris. Others that will go unnamed whose assistance and contributions are most valued include representatives from the U.S. intelligence community, representatives of international oil and gas companies, both active and retired, former U.S. Ambassadors to one or more of the countries under study, desk officers at the U.S. Department of State, members of the Office of the Secretary of Defense who follow developments in these countries, representatives of those countries' missions to the United States, members of a variety of think tanks that follow international energy topics, and academics who have published articles on energy topics. I am grateful to them all. Their knowledge and their willingness to offer and share it contributed greatly to my understanding. Some indicated a desire to remain anonymous, and I have adhered to their wishes. Finally, I respect and appreciate the efforts of my readers of the drafts and I express my appreciation to the holders of copyrighted material for their willingness to release it. All remaining errors are mine alone.

Abbreviations

ABTC. Aktau–BTC proposed pipeline link

ACG. Azeri–Chirag–Guneshli

ADB. Asian Development Bank

Agip KCO. Agip Kazakhstan North Caspian Operating Company

AIOC. Azerbaijan International Operating Company

ANB. Azerbaijan National Bank

API. American Petroleum Institute, which sets the degree-gravity standards for measuring oil quality

bcm. billion cubic meters

bcm/y. billion cubic meters per year

BOTAS. Turkey's state natural gas and pipeline company

BTC. Baku–Tbilisi–Ceyhan Company and/or BTC pipeline

CAC. Central Asia–Centre (CAC) pipeline, in fact not a single line but a network of five separate lines, commissioned between 1966 and 1987. CAC-3 runs through Kazakhstan, and CAC-1, 2, 4, and 5 pass through Uzbekistan

CERA. Cambridge Energy Research Associates

CIA. Central Intelligence Agency

CITES. Convention on International Trade in Endangered Species of Wild Fauna and Flora

CNODC. China National Oil and Gas Exploration and Development Corp.

CNOOC. China National Offshore Oil Corporation

CNPC. China National Petroleum Corporation

CPC. Caspian Pipeline Consortium

DDT. dichloro-diphenyl-trichloroethane

DOE. U.S. Department of Energy

DOE/EIA. U.S. Department of Energy, Energy Information Administration

DRAs. drag reduction agents, used to speed the flow of oil through a pipeline, and hence, the volume per year

EBRD. European Bank for Reconstruction and Development

EITI. Extractive Industries Transparency Initiative

GDP. gross domestic product

IAEA. International Atomic Energy Agency

IEA. International Energy Agency

ILSA. Iran–Libya Sanctions Act

IMF. International Monetary Fund

KIO. Karachaganak Integrated Organization

LNG. liquefied natural gas

MMBtu. million British thermal units

mmb/d. million barrels per day

NBK. National Bank of Kazakhstan

NICO. Naftiran Intertrade Co. Ltd.

OECD. Organization for European Cooperation and Development

OGJ. Oil & Gas Journal

OSCE. Organization for Security and Cooperation in Europe

PSA. Production sharing agreement

PWYP. Publish What You Pay

RFE/RL. Radio Free Europe/Radio Liberty

SCP. South Caucasus Pipeline

Sinochem. China National Chemicals Imports and Exports Corporation

Sinopec. China Petroleum and Chemical Corporation

SOCAR. State Oil Company of the Azerbaijan Republic

SOFAZ. State Oil Fund of Azerbaijan

TCO. Tengizchevroil

TIPline. Trans-Israeli Pipeline

TPAO. Turkish Petroleum Corp.

UNEP. United Nations Environment Programme

WTO. World Trade Organization

1

Introduction

Ever since the demise of the Soviet Union, energy pundits and Western government representatives alike have touted the oil and gas riches of the Caspian region. The countries typically included as "inner circle" and potentially major oil and gas producers in the region are, in addition to Russia and Iran, Azerbaijan west of the Caspian Sea, together with Kazakhstan and Turkmenistan to the east. Georgia is typically included as an inner circle major transit country for oil and gas pipeline shipments. Uzbekistan is generally excluded from consideration as a member of the inner circle since it is not a littoral Caspian country. We include it in our outer circle of interested countries both because of its reserves and its role in energy transportation. In all the years since the 1991 breakup of the Soviet Union and the newfound independence of these former Soviet states, very little in the way of additional oil and gas supplies has reached international markets, despite billions of dollars in investment and much hype in the general and oil industry press. The reserves are not as great as once thought, and are nowhere near the 200 billion barrels some suggested initially. Rather, the US government places them at 17–33 billion barrels, although BP's 2005 estimates put them at 47.7 billion barrels. In addition, internal politics, technical impediments, and commercial obstacles continue to delay and limit the prospects for production from the region.

The thesis of this paper is simple. It is that both oil reserves and future production levels have been overhyped and overestimated. The dreams have faded as the hard realities of energy development and politics have set in. We project that future Caspian oil production will be about 3.2 million barrels per day (mmb/d) in 2010, rise to 3.6 mmb/d in 2015, and decline to about 3.3 mmb/d in 2020. This projection includes accounting for the inevitable technical and political delays affecting production and export, and incorporates the view that whatever target production levels are, production will reach world markets far later than proponents of the Caspian's potential would have us believe. The goal of Kazakhstan's president, for example, is for that country alone to produce 3.5 mmb/d by 2015; we believe this is unlikely to occur. The difficult political, transport, and technical environments have not been adequately factored in, with the result that one can anticipate continuing delays and postponements of target production levels. Profitability remains an issue, especially if oil prices drop. Nor has the size of international gas markets been properly assessed in terms of risks to both producers and consumers. The theme of enthusiasm for the resources continues politically, even as energy firms have come to recognize that this is a high-cost region and a politically difficult and highly unstable arena.[1] While the region will certainly provide the world with some future energy supplies, and while oil is highly fungible, these supplies have been overestimated in terms of size and timing, and underestimated in terms of both political and economic risks. Moreover, these Caspian volumes are likely to be swamped by the large amounts of oil presently under development and likely to hit the market in the future at about the same time.

This paper examines the countries surrounding the Caspian Sea for their potential as oil and gas producers and contributors to world energy security and diversity over the period to 2015. The focus is primarily on oil since it is a fungible product and can be transported to anywhere in the world once it reaches a seaport. In the future, gas markets may develop, but their time frame is longer-term. The early enthusiasm for the Caspian in large part reflects the desires of consuming nations to have an alternative to Middle East oil supplies, given all the uncertainty and instability in the expected production and exports from that region. But at best, in our view, the Caspian region will contribute only roughly 3 to 3.6 percent of world oil use in the future.

The paper also considers the stakes involved, including forecast revenues and their potential uses, and issues of governance affecting the disposition of these revenue streams. In addition there are other issues, such as the delimitation of the Caspian seabed and waters among the littoral states, several environmental considerations, the expected nature of the local leadership, and the prospects for either continuing autocracy and repression or nascent democracy. The stakes are high for the international energy companies that have invested in the region. Billions of dollars have been committed, but the political and business risks, as well as the geological and transportation challenges are enormous. Over time, the investment climate has noticeably worsened as states try to increase their shares of the economic rents.

For the most part these nations are authoritarian states, with no checks on the power of the ruler, despite the countries' adoption of democracy in name at the time of independence. Elections are regularly rigged and are sham events, as demonstrated most recently in the polling in both Azerbaijan and Kazakhstan. More precisely, these states are "one-bullet regimes," as one of my sources characterized them, in that most would be plunged into disarray or even civil war were their authoritarian leaders removed from office. The level of discontent is high in virtually all of these countries. After the experience of a Rose Revolution in Georgia and an Orange one in Ukraine, it is clear that these states are indeed fragile and in many cases are tinderboxes waiting to ignite. These political "flower" developments sent major tremors through the governments of Kazakhstan and Uzbekistan, although not of Turkmenistan, where information about the outside world is more vigorously filtered and controlled. The iron hand of Uzbekistan's ruler was challenged again in 2005, when a popular uprising in the Fergana Valley was put down with brutal force and much loss of life. The future of these states will play out differently, whether or not the borders of the states as we now know them remain intact. Not only do the individual countries distrust each other, despite calls from other interested parties for regional cooperation, but each also has differing attitudes toward either cooperation or coexistence with the more powerful regional neighbors. These countries are for the most part kleptocracies governed corruptly, and are abjectly poor. All face a plethora of political and economic problems that oil revenues may or may not be used to resolve. Several do not merit the moniker of state; in fact they are failed states, victimized by their current and past leaders.

Outside the countries themselves, one must look as well at the ever-changing attitudes and energy and foreign policy interests of the near neighbors, Turkey and China, and of those of the distant United States. Oil and gas economics are inextricably interwoven with domestic political and ethnic issues in the Caspian, and are likewise interwoven with politics in the more powerful states. Moreover, the United States has for some time altered the playing field with its military presence in the region. Some years ahead, the enthusiasm about the promise of the Caspian region may indeed wane, as prospects in the much more generously endowed oil-producing countries of Saudi Arabia, Iran, and Iraq, and those in Africa and elsewhere continue to dwarf the Caspian players on the world petroleum stage. This book addresses the issue as to whether the Caspian as an oil and gas play really matters in considerations of formulating current U.S. energy policy, and we conclude that it does not. Moreover, on national security grounds, we conclude that it was a mistake in both foreign and energy policy for the United States to have pushed for the BTC and CPC pipelines, as well as for the Murmansk pipeline project. We likewise conclude that the Caspian region is not of significant strategic energy interest to the United States.

Chapter 2 looks at a number of prospects and problems specific to energy developments in the region. It considers the changing views on energy reserves, historical and potential production of oil and gas, energy security aspects, transport alternatives, costs and prices, issues preventing agreement on delimitation of the seabed, environmental concerns, Dutch Disease/deindustrialization issues resulting from a resource boom, and the (limited) prospects for greater transparency and use of oil revenues. Azerbaijan's future is linked to its exploitation of oil reserves discovered in the Soviet era, and a single large new gas field discovered thereafter. Exploration to date has been disappointing in that no major new Azeri oil or gas fields have been discovered. Central Asia is primarily a natural gas province, and suffers from the dearth of developed gas markets other than those served by Soviet-era pipelines passing through Russia. Oil discoveries in Kazakhstan, however, suggest that the North Caspian has some promise for future oil production. We find that forecast oil output for the region is likely to be far lower than others predict, and the risks of both political battles between companies and governments and of future Dutch Disease and misuse of revenues in the producing countries are great.

Chapter 3 examines what we term the inner circle of Caspian energy producers and transporters, and the markets they serve. The inner circle is composed of Azerbaijan, Georgia (even though it is not a Caspian littoral state), Kazakhstan, Turkmenistan, Iran, and Russia, and we examine the Caspian energy resources of each. We include Turkey as a de facto member of the inner circle, for its geographic proximity is key in plans to move Caspian energy supplies to Western markets. We find a number of problems of an economic, technical, and political nature affecting both the Caspian members of the inner circle and Turkey. In Azerbaijan, only two big fields have been found: the offshore ACG oil field, and the offshore Shah Deniz natural gas field. The country has been a producer for a hundred years, and its onshore fields are in sharp decline. Many fields licensed after the country's independence have been shut down or abandoned because commercial prospects were not discovered. ACG production will mount sharply to about 1–1.2 mmb/d by 2010, now that the BTC pipeline is in operation, but production will decline nearly just as precipitously in the years thereafter. Gas markets beyond Turkey are promising, but there are many uncertainties surrounding future sales to Western Europe, in addition to the price competition Caspian gas will face from Russian gas. Oil and gas output are risky owing to geology and politics, where the latter includes not only the government's taxes and other terms, but also the risks of renewed hostilities on the Nagorno-Karabakh issue possibly affecting pipeline operation.

Georgia plays an intermediary role, since both the BTC pipeline and its South Caucasus Pipeline for natural gas pass through the country. It is counting on transit fees to give its budget a big boost and permit it to diversify its energy supplies, but those outcomes will clearly depend on the successful operation of both lines and the absence of interruptions from either terror groups or its restive breakaway provinces.

To the east of the Caspian Sea, Kazakhstan's three big oil and gas fields are Tengiz, Karachaganak, and Kashagan. All have faced slipping timetables and battles with the government as the investment climate has worsened for the companies involved. The government also launched what can only be called an assault on the Canadian company PetroKazakhstan, leading to its sale to CNPC (and a piece of PetroKazakhstan sold to KazMunaiGaz). This assault has further soured the investment environment. The three major fields are promising but costly to develop and to arrange transportation. While Kashagan has yet to begin oil production, the guessing game continues as to whether its promise will merit a new crude oil—dedicated export pipeline. Our view is that it will not, for there will be space on BTC and other lines. Nonetheless, President Nazarbayev is keeping transportation options open in all directions, and moving ahead with plans for a trans-Kazakhstan oil pipeline to China.

Turkmenistan and Uzbekistan suffer from politics inappropriate for energy development. While Turkmenistan is rich in oil and gas resources, there is little foreign private investment interest there, given the political regimes' idiosyncrasies and economic controls. Turkmenistan's gas reserves are substantial, but President Niyazov is viewed as neither predictable nor dependable. He failed to come to terms with Azerbaijan on the proposed subsea oil and gas pipelines. Niyazov's gas sales to Russia and Ukraine and his hopes of selling additional large quantities to Pakistan and India through the possibly revived trans-Afghan gas pipeline project may have overcommitted Turkmenistan's gas reserves.

Iran has yet to start drilling for either oil or gas in the Caspian, despite its announced intention to do so, perhaps because its reserves are so extensive elsewhere in the country. It is nonetheless assuming a growing role as a transit country in terms of volumes of Russian and Caspian oil swapped in the north for the value of equivalent barrels delivered in the south. Iran's efforts to sell gas to Pakistan and India may lessen the chances of Turkmenistan gas sales to those two countries. Iran's intransigence, along with the intransigence of Turkmenistan, on Sea delimitation poses a problem for future production and transport options. U.S. sanctions on Iran and the continuing differences between Iran and the West on its nuclear program have to date lessened the chances of a future oil export pipeline carrying Caspian oil to pass through Iran to the Persian Gulf.

Russia is one of the world's largest oil and gas producers in recent years without any Caspian production to speak of, but it is now preparing to develop fields that straddle the Sea's median line of national sector delimitation. Its tax and investment regime, as a result of changes in the rules and the Yukos affair, temporarily slowed private oil company interest in Russian oil development, including in the Caspian. Persistent high world oil prices, however, have worked to overcome foreign investor skepticism, although risks remain high. Russia's support of the monopoly transportation companies, Gazprom and Transneft, suggest that the energy sector will remain firmly in the hands of the state. Russian oil output has grown by 50 percent since 1996, and the country has repeatedly rejected the overtures of OPEC to join the cartel. In a period of relatively high oil prices, we see very little Russian interest in cooperating with the United States on oil matters under the Energy Dialogue begun in 2002, but Russia is seeking international cooperation and assistance from the United States and others on development of natural gas and LNG export projects. While Russian oil production growth is not likely to proceed at as fast a rate in the future as it did in recent years, Russia has targeted an output level of 12 mmb/d by 2012. Russia, together with Kazakhstan, appears well positioned to meet large fractions of China's future energy demand.

Chapter 4 addresses what we call the outer circle of countries that have interests in Caspian energy developments. These include China, Uzbekistan, and the United States. While Pakistan and India may also have energy interests in the Caspian region, we have included them only tangentially, in terms of discussions of proposed pipeline routes. China is looking for oil and gas throughout the world, and the Caspian is no exception. Should China's economic growth slow, its projected future demands for oil and gas will decline as well, but it will still seek to enhance its energy security by continuing the diversification of its imports away from the Middle East. The proposed oil pipeline spur from Russia to Daqing has long been a hope for China, but for Russia it does not seem to make much sense, since oil sent to Daqing would face no competition for buyers. The alternative, however, is an oil pipeline from Russia to Nakhodka on the Russian Far East coast, from where oil can go anywhere on tankers from all over the world. Russia has vacillated on its export plans to the east; it now appears that it will build both links, although the timetable is uncertain. In addition, the oil export line from Atasu in central Kazakhstan to China is now finished, and a gas export line could follow some time in the future, provided that natural gas imports by pipeline can be delivered at prices competitive with Chinese domestic coal.

Uzbekistan, under its autocratic leader, is considered to be of no energy or security significance. Its only significance to energy trade and transport are the oil and gas pipelines from the Soviet era that traverse its territory. While it has reserves of both oil and gas that might attract Western foreign investment, it is clear the country is not interested in abandoning its central planning and liberalizing its economy to bring about such interest. This is a state that is at severe risk of either implosion or explosion, given its autocratic leadership and the abject poverty and discontent of its populace. The leadership vehemently represses internal criticism and rejects external criticism, as the United States discovered after its criticism of the Andijan massacre and its subsequent eviction from its leased Uzbek airbase at Karshi-Khanabad.

The United States' energy interest is related to new, non-Middle East supplies of oil and gas to improve world energy security. This interest has been eclipsed, however, by its determination to make its presence felt in the battle against terrorism, even at the cost of walking away from its commitments to democracy and prosperity as it supports despotic regional leaders. We find that energy issues and developments influence all the great powers' interests, but that the strategic energy interests of the United States in this part of the world are minimal. Our examination of U.S. national interests leads us to conclude that we have very few in the area, particularly as related to energy. For all its interest in energy security, the United States has failed to recognize that this region's promise is really quite small. Likewise, for all its interest in energy security, the United States has pursued its course in the region more from a foreign policy perspective to disadvantage both Russia and Iran than to assist the newly independent states and foster their stability under democracy. Presently, the interest in waging war on terrorism has overshadowed what until that time was the overriding U.S. interest in new energy supplies and enhanced world energy security.

Chapter 5 offers our conclusions. We find that the Caspian region as an oil and gas play has been much overestimated in terms of both reserves and actual or future production size, and is unlikely to make much of a difference to future world energy supplies, particularly in comparison to the Middle East. As much as 20 mmb/d or more of new oil and liquids will be arriving on the world market between now and 2015, or 16 mmb/d by 2010, of which Caspian supplies are only a small portion. Gas supplies worldwide are on the rise, limited only by demand and transportation. In addition to questioning the hype that still surrounds the Caspian as an oil region (although diminished from its earlier levels), we conclude that political problems and technical issues are at least as important as oil and gas reserves. This is a region that is very difficult to do business in. It is landlocked, the geologic conditions are demanding, production is costly and subject to repeated slips in schedules, and the politics are autocratic, unpredictable, and Byzantine. The chances for democracy are slim, and failed states are present not only now but likely to be so in the future. The risks of continued inept and bad governance, along with the onset of Dutch Disease, are great, and we see no indications that the countries will avoid the problems associated with managing significant revenue inflows. In addition, the continuing poverty despite the inflows of oil and gas revenues is likely to lead to more unrest in the years ahead, particularly under authoritarian leaders. In short, we find this region to be one whose governance does not live up to its potential as an oil and gas producer. Moreover, that potential itself is small compared to the world outlook for future oil supplies.

2

Caspian Energy: Prospects, Pipelines, and Problems

In this chapter we address the size of oil and gas reserves, current and anticipated production, world energy security, issues of transportation and risk, costs, the continuing problem of Caspian Sea delimitation, environmental issues, and anticipated revenues to governments as either keys to prosperity or the "resource curse." That curse, otherwise known as Dutch Disease (first observed in the Netherlands after significant finds of natural gas in the 1960s), occurs when exploitation of natural resources raises the value of the local currency, thereby discouraging exports and increasing imports, and simultaneously leads to the deindustrialization of the local economy. We look at the evidence on Dutch Disease in both Azerbaijan and Kazakhstan, and consider alternative means of mitigating it. The focus is primarily on oil, for that is the fungible commodity in key demand in world markets; outlets for these countries' natural gas remain limited at the present time.[1]

From independence until about 1994, the United States paid little attention to these former Soviet republics, with the exception of Kazakhstan, which had a large concentration of former Soviet Union nuclear weapons. Following the first war in Iraq, the United States remained interested in worldwide energy developments, but treated Southern Caucasus and Central Asia as "exotica."[2] Until that time, U.S. interest was shown in the Freedom Support Act of 1992, which assisted former Soviet republics but restricted U.S. assistance to Azerbaijan. Baku and Yerevan were at war at the time, and for that matter still are, despite the existence of a shaky 12-year ceasefire, over the Azeri province of Nagorno-Karabakh. Azerbaijan had cut off natural gas deliveries to Armenia. Although international oil companies were aware of the possibility of undeveloped or undiscovered energy reserves in the Caspian Sea region since before the collapse of the Soviet Union, the U.S. government did not focus its attention on the area until the run-up to the signing of the 1994 "Contract of the Century" between the government of Azerbaijan and the international oil consortium AIOC.[3] Azerbaijan's president was feted in Washington after the signing of the contract to develop the offshore Caspian oilfields known since Soviet times and referred to as the ACG fields. Across the Caspian Sea in Kazakhstan, Chevron (now ChevronTexaco) began development of the large onshore Tengiz field in 1993.[4] The signing of both these contracts, together with the Russian invasion of Chechnya in December of 1994, heightened U.S. attention on what had been a relatively obscure area of the world and one beset with ethnic and political conflicts. Both energy projects were landlocked, and would need pipeline connections either to the Black Sea, to the Mediterranean, or to the Persian Gulf; early volumes of oil moved largely by rail or through the existing Russian and Georgian pipelines until new pipelines could be agreed and commissioned.

ENERGY RESERVES

The mythology on the richness of the Caspian's energy resources grew rapidly over the years following the mid-1990s. Some estimates placed proven or recoverable reserves as high as 200 billion barrels of oil. The area rapidly came to be called a "strategic" region because of these optimistic energy numbers. Little distinction was made between estimated oil in place, proved reserves, probable reserves, and possible reserves. Enthusiasts were prone to likening the area to another Middle East, although that region's oil reserves at the time were placed at about 676 billion barrels.[5] Current estimates of Caspian-region proved oil reserves are far below those of the Middle East.

Among the earliest reports alleging the strategic significance of the region was former National Security Council staffer Rosemarie Forsythe's 1996 thoughtful paper entitled "The Politics of Oil in the Caucasus and Central Asia."[6] Focusing on the Caucasus and Central Asian states as actual and potential oil producers of "considerable geo-strategic importance"[7] for the future, Forsythe pointed out that the largest oil deposits are in Azerbaijan and Kazakhstan. This remains true. The Soviets had estimated the reserves of the region at approximately10 billion barrels,[8] which would by size alone not justify calling the region strategic in terms of world oil reserves of nearly 1,300 billion barrels. Summarizing the history of Azerbaijan's oil production from the 1870s to its peak in the 1940s, Forsythe noted the fall in Azeri oil production as the Soviet Union shifted its interest from the Azerbaijan fields to those in Western Siberia. In Kazakhstan as well, oil production has a long history, with the Soviets exploiting oil deposits in Kazakh basins since the 1950s. Yet the new reserves estimates circulating in the 1990s far exceeded the former Soviet estimate and were wishfully interpreted to portray the Caspian as a major new alternative to Middle Eastern or Persian Gulf oil supplies.

In May of 1997, the U.S. Department of State added to the reserves euphoria.[9] It stated that potential, including undiscovered and possible, reserves could be as much as 200 billion barrels of oil and nearly 8 trillion cubic meters of gas.[10] Since 1997, reality has set in, and reserves estimates have been defined more carefully. Proved reserves are quantities that are known with 90 percent certainty in terms of geology, prevailing prices, and the state of technology. The U.S. Department of Energy's independent statistical arm, EIA, reports Caspian region oil reserves at 17–33 billion barrels, far below the euphoric estimates of years gone by, yet slightly less than BP's 2005 estimate.[11] Proved reserves will increase or decrease over time as more drilling takes place—which adds to the body of knowledge about the reservoirs—as new techniques and technology are brought to bear in the exploration and production process, or as prices rise to make more fields commercially viable.

Often compared erroneously with the North Sea's petroleum reserves, the Caspian region has yet to rival the former's original size as a reserves basin. The argument has been made that the optimistic and elevated levels of Caspian reserves were purposefully but wrongly promulgated to assuage the concerns about rising U.S. energy consumption, diminishing domestic production, and growing import dependence. According to this explanation, if such high levels of untapped Caspian resources were truly available and open to exploitation, then American consumers would not need to worry about potentially higher prices or scarcities in the future (except in special cases like the war in Iraq or in the aftermath of U.S. Gulf Coast hurricanes), and the United States would not need to feel pressed to cut energy consumption. If accessible and seemingly limitless oil supplies were located in the Caspian, the argument went, then the energy well-being of American and Western consumers was virtually ensured for years to come.[12]

Sense and truth have a way of ultimately displacing invention and fantasy. Three respected sources for worldwide proved oil and gas reserves are DOE/EIA, BP, and the *Oil & Gas Journal* (*OGJ*). Table 2.1 suggests that these three sources are now close to agreement about the size of proved oil and gas reserves in Azerbaijan, Kazakhstan, Turkmenistan, and Uzbekistan, and that they amount to only a fraction of the 200 billion barrels previously estimated.

The more recent data include the July 2000 discovery of the offshore Kashagan oil and gas field in Kazakhstan. This field was declared commercial a year later, and may begin production in 2008, a target date that has already slipped nearly 3 years from the original target date of 2005. EIA estimates oil reserves for this field alone at 7–9 billion barrels, with potential reserves, if

Table 2.1. Estimated Proven Oil and Gas Reserves in Key Caspian Region Countries (Billion barrels of oil or trillion cubic meters of gas)

		Azerbaijan	Kazakhstan	Turkmenistan	Uzbekistan	Total
BP 2005[13]	Oil	7.0	39.6	0.5	0.6	47.7
	Gas	1.37	3.0	2.90	1.86	10.83
OGJ 12/2005[14]	Oil	7.0	9.0	0.5	0.6	17.1
	Gas	0.84	1.82	2.0	1.9	6.6
BP 2004[15]	Oil	7.0	9.0	0.5	0.6	17.1
	Gas	1.37	1.90	2.90	1.85	8.02
OGJ 12/2004[16]	Oil	7.0	9.0	0.5	0.6	17.1
	Gas	0.84	1.82	2.0	1.9	6.6
EIA 8/2003[17]	Oil	7–12.5	9–17.6	0.5–1.7	0.3–0.6	17.2–32.8
	Gas*	0.84	1.82	2.0	1.8	6.5
Petroconsultants 1998[18]	Oil	12.5	17.6	1.7	0.3	32.1
IEA 1998[19]	Oil	3–11	22	1.5–47	n/a	15–40
	Gas	0.3–0.8	1.5–2.4	2.7–21	n/a	6.7–9.2

Other estimates of proved oil and gas reserves (billion barrels, or trillion cubic meters) for region, not by country:

Estimates published in 2002	18–35 billion bls[20]
	18–34 billion bls; maybe maximum 70 billion bls[21]
Estimates published in 2000	25–30 billion bls; ultimate 40–60 billion bls[22]
	15–200 billion bls; 6.4–7 tcm[23]
	At least 30 billion bls[24]
	15–40 billion bls; ultimate 70–150 billion bls[25]
	15–40 billion bls; 6.4–6.5 tcm[26]
	15–31 billion bls; ultimate 60–140 billion bls[27]
	15–35 billion bls; ultimate 100–200 billion bls[28]
Estimates published in 1999	15.6 billion bls, 163 billion bls possible; 8.3 tcm gas, 9.3 tcm possible[29]
	20–25 billion bls; 50–75 billion bls proved and possible[30]
	15–30 billion bls[31]
	15–31 billion bls; 6.5–10.2 tcm[32]
Estimates published in 1998	20–35 billion bls[33]

* Data converted to cubic meters from cubic feet.

Source: BP, *Statistical Review of World Energy*, June 2004 and June 2005; Anonymous, "Worldwide Look at Reserves and Production," *Oil & Gas Journal*, Vol. 100, Iss. 52, December 23, 2002, 114–115, and Vol. 102, Iss. 47, December 20, 2004, 22–23; DOE/EIA, "Caspian Sea Region: Key Oil and Gas Statistics," August 2003, at http://www.eia.doe.gov/emeu/cabs/capstats.html, as cited in Department of Energy, Energy Information Agency, *International Energy Outlook 1998*, 34; and International Energy Agency in Co-operation with the Energy Charter Secretariat, *Caspian Oil and Gas: The Potential of Central Asia and Transcaucasia* (Paris, France: OECD/IEA, 1998), 32.

secondary recovery techniques are used, of as much as 9–13 billion barrels.[34] The size of natural gas reserves is not the key determinant for exploitation, since gas markets outside the former Soviet Union have yet to develop. Gas reserves estimates are of consequence, however, for Azerbaijan, which plans to deliver gas via pipeline to Turkey and, in the future, to other European customers, before the other Caspian producing countries are able to develop new Western markets for their natural gas.

If one accepts the reserves estimates of 17–33 billion barrels of oil for the Caspian region, one then needs to look at these reserves relative to the world at large. BP estimates end-2004 world proved oil reserves at 1,189 billion barrels, while the *OGJ* gives January 2006 world reserves of 1,293 billion barrels.[35] At the most, the proved oil reserves of the Caspian region amount to 1.3–2.8 percent of the world's total. By comparison, oil reserves in Saudi Arabia alone amount to 22 percent of the world's total. Natural gas reserves, however, are substantial, and amount to 6 percent of the world's total, but they are not in the same league as those of either Russia or Iran, which are, respectively, more than 27 percent and more than 15 percent of the world's total.

OIL AND GAS PRODUCTION

Just as there have been differences in Caspian oil and gas reserves estimates, there are differences in estimates of how much oil and gas production will occur, and when. These estimates are integrally tied to demand estimates; to the timing, size, and costs of transport options to move the oil and gas to markets; to the availability of drilling equipment; and to host-government policy and fiscal decisions. As landlocked states, these producing countries face major transportation challenges involving third-party countries.

The Caucasus and Central Asian economies for the most part have little else to market, and each actual or potential hydrocarbon producer and exporter hopes for substantial revenues.[36] Each country has a high discount rate and wants revenues now rather than later, with the result that each wants foreign developers and state oil companies to produce as much as possible as soon as possible. Countries seek and win signing bonuses ahead of production, and bonuses when production reaches certain milestones; bonuses are ways to obtain revenue sooner rather than later. All see their future production far in excess of domestic consumption, with the difference to be exported and sold at world prices, provided there is transport to reach world markets.

Historical Production

Oil production in 2005 from the Caspian region was approximately 2 mmb/d, or about 2.8 percent of world oil production.[37] In the past nearly

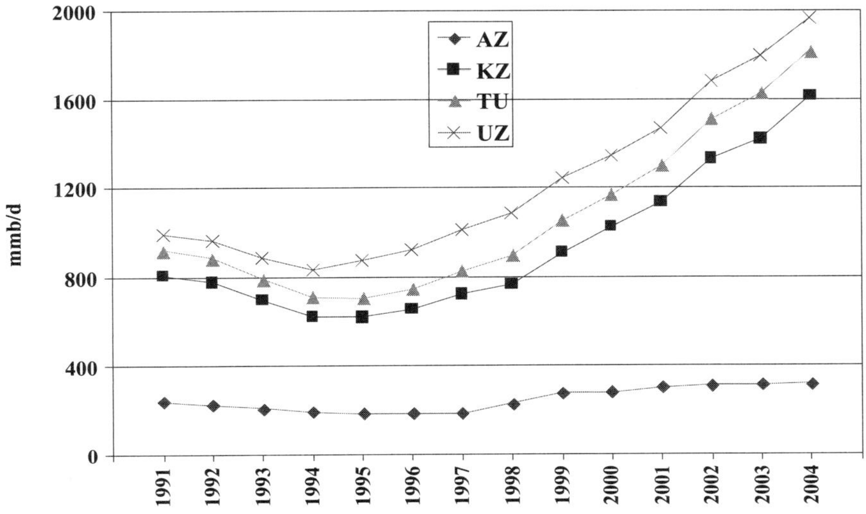

Figure 2.1. Caspian Oil Production, 1991–2004
Source: BP, *Statistical Review of World Energy*, June 2003, June 2004, and June 2005.

all moved through Russia, with the exception of one pipeline carrying oil from Baku to Supsa on the Black Sea and some swaps through Iran. With the opening of BTC in 2006, however, large new volumes will cross Azerbaijan and Georgia to reach Turkey. Figure 2.1 shows historical Caspian oil production.

The recovery in Caspian oil production following the Soviet collapse and the low levels of the mid-1990s is quite remarkable, as shown in Figure 2.1. Most current production comes from Azerbaijan's ACG fields, Kazakhstan's Tengiz field, and, within a few years, from Kazakhstan's Kashagan and Karachaganak fields. Some older fields, whose output is declining, still produce in both countries. In Azerbaijan, AIOC began exporting oil by pipeline in 1998 through Russia and Georgia, after making improvements in the existing pipeline to Supsa on the Black Sea. Thereafter oil reaches international markets through the Bosporus. The figure shows that production in Azerbaijan has risen more than 90 percent since 1997. Kazakhstan's 2005 production was more than 1.2 mmb/d. Production rose with the onset of shipments of Tengiz oil a few years ago, first by rail and then by pipeline. Since the 2001 completion of the CPC 1,580-km line from Tengiz across Kazakhstan and Russia to the Black Sea for tanker shipment to world markets, Tengiz output has been rising. CPC plans to expand design capacity to 1.3 mmb/d.[38] Both governments and oil companies recognize that it is not economic to build up producing capacity if there are no ways of moving the output to market.

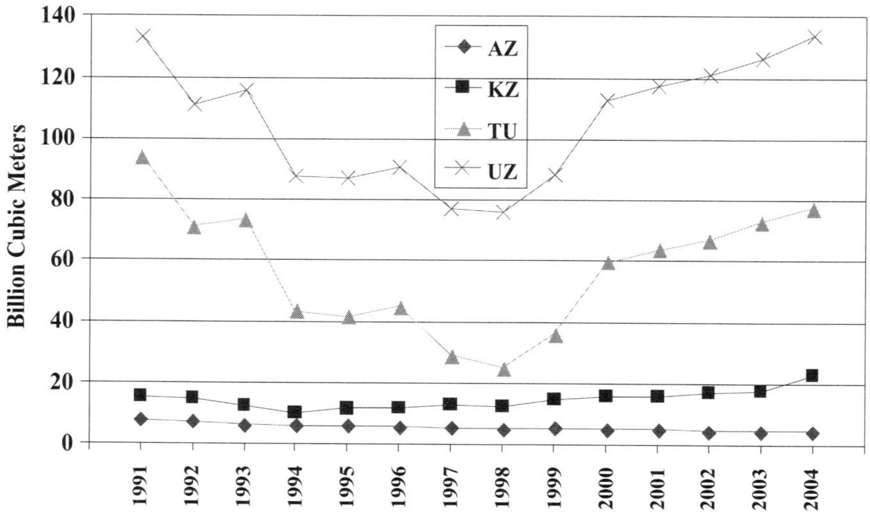

Figure 2.2. Caspian Natural Gas Production, 1991–2004
Source: BP, *Statistical Review of World Energy*, June 2003, June 2004, and June 2005.

Turkmenistan's and Uzbekistan's oil production is largely consumed domestically, although Turkmenistan and Kazakhstan swap some oil with Iran, which sells an equivalent number of barrels at the Persian Gulf on their behalf. Turkmenistan's export and pipeline regulations, its state monopoly buyer that pays producers less than world market prices, and the leader's mercurial character have to date stalled foreign interest in undertaking oil projects there. Uzbekistan more than doubled its oil production between 1992 and 1999. While currently a net oil exporter, the country must replace the old fields' output with new production. Tashkent is campaigning to attract foreign investors to expand production at both existing and new fields, although political unrest, an unreformed economy, and absence of pipelines transiting countries other than Russia are deterrents. With the exception of Chinese state oil companies, however, foreign private investor interest focuses particularly on the production prospects in Azerbaijan and Kazakhstan; neither the resources nor the politics are attractive in the other two countries.

As regards natural gas production, these four countries produced 134 bcm in 2004 (2005 figures for gas production are not yet available), or about 5 percent of world gas output (Figure 2.2). The fifth country, Georgia, is not a natural gas producer, but rather a future gas transit country for Caspian gas destined for Turkey and other Western markets. Natural gas export markets and transportation links in this region are far less developed than those for oil. Despite the size of the countries' natural gas resources, all gas pipeline proposals confront political risks, the difficulties of transiting neighboring countries, large capital requirements, and the need to win long-term buyers who can be depended on to meet these countries' contractual payment obligations at world market prices.

Historically a net importer of gas, Azerbaijan is developing its large Shah Deniz natural gas field. Plans call for using some of this gas locally to replace Russian imports as well as for selling 6.6 bcm/y to Turkey, and perhaps additional volumes to Western Europe if new gas pipelines are built. Kazakhstan's gas resources have yet to find a market, although there have been repeated unrealized reports of a future natural gas line eastward to China. Kazakhstan is likely to use some of its gas for reinjection to maintain reservoir pressure for oil production, and to sell some to Russia. After Russia, Uzbekistan is the leading natural gas producer among the republics of the former Soviet Union. Uzbekistan is attempting to speed up development of both existing and new fields, largely for domestic consumption but possibly for future export to China. Turkmenistan has the largest gas deposits of the four former Soviet republics, but saw its output diminish when Russia in 1993 ended an agreement to credit Turkmenistan's gas deliveries to Russia with equal amounts of Russian gas exports to Europe. Turkmenistan thereafter negotiated future gas sales agreements with both Ukraine and Russia, and is seeking new sales to Pakistan and India. Questions remain on whether existing and potential contracts may overcommit Turkmen gas reserves.

Potential Production

Still in question is what current or future revised reserves figures could support in terms of projected oil and gas production. Moreover, production destined primarily for export will serve an uncertain level of world oil and gas demand at unknowable future prices. Estimates of future oil production vary greatly, according to the source and the date of the estimate. The future supply of gas will be constrained by the development of markets and the expansion of paying customers.

An alternative approach to estimating future production is the examination of output profiles for major field development projects currently or soon to be underway. Four projects account for some 70 percent or more of regional oil reserves. Five fields will drive future oil and gas output and account for most of the capital expenditures on development. These include the offshore ACG oil fields and the offshore Shah Deniz gas field in Azerbaijan, the Tengiz and Karachaganak onshore oil fields in Kazakhstan, and the large Kashagan oil field in Kazakhstan's offshore waters. While other

Table 2.2. Projections of Future Caspian Oil Production ('000 b/d)

		2010	2015	2020
Azerbaijan (AZ)	Azeri-Chirag-Guneshli (ACG)	1,000	700	380
Kazakhstan (KZ)	Tengiz	700	750	750
	Karachaganak	240	300	225
	Kashagan	450	1,200	1,200
	Other KZ*	300	400	400
	Total KZ	1,690	2,650	2,575
Total KZ & AZ—key fields		2,690	3,350	2,955
Uzbekistan (UZ)		200	200	200
Turkmenistan (TU)		200	200	200
Russia and Iran (Caspian area)		Negligible	Negligible	Negligible
TOTAL		**3,090**	**3,750**	**3,355**

* Estimate

Source: ACG from AIOC development plan (see Figure 2.3); Tengiz and Karachaganak estimated production from press reporting; Kashagan from Nick Griffin and Caroline Cook, Kashagan – Catch of the day? Deutsche Bank, February 25, 2004, 2, Private Report. Uzbekistan and Turkmenistan estimated, based on M. R. Farzanegan, "Iranian options most economically viable for exporting Caspian oil," *Oil & Gas Journal*, Vol. 101, Iss. 11, March 17, 2003, 22, and interviews with industry sources.

projects and prospects exist, particularly in Kazakhstan, they are not likely to make a major impact on regional production in the next 10–15 years. Moreover, old onshore production in Azerbaijan is declining, and new large Azeri fields have not been found. Table 2.2 shows our best and current estimate of future Caspian oil output.

The Kazakh government has announced—perhaps with some measure of wishful thinking not linked to practical planning—that Kazakh oil production alone will amount to 2.3 mmb/d in 2010 and 3.5 mmb/d in 2015, but these are levels that a number of international oil companies have publicly doubted.[39] After 100 years of production in Azerbaijan, there is some doubt in the minds of experts whether there are any yet-undiscovered prospects. In terms of geology, some observers conclude that the only viable prospects are on the Azeri Apsheron trend, with everything north of this area outside the promising hydrocarbon zone, and everything south of this area a deep gas zone extremely difficult to drill.[40] The drop in ACG production after about 2010, as shown in Figure 2.3, is unlikely to be offset by substantial new finds, either onshore or offshore.

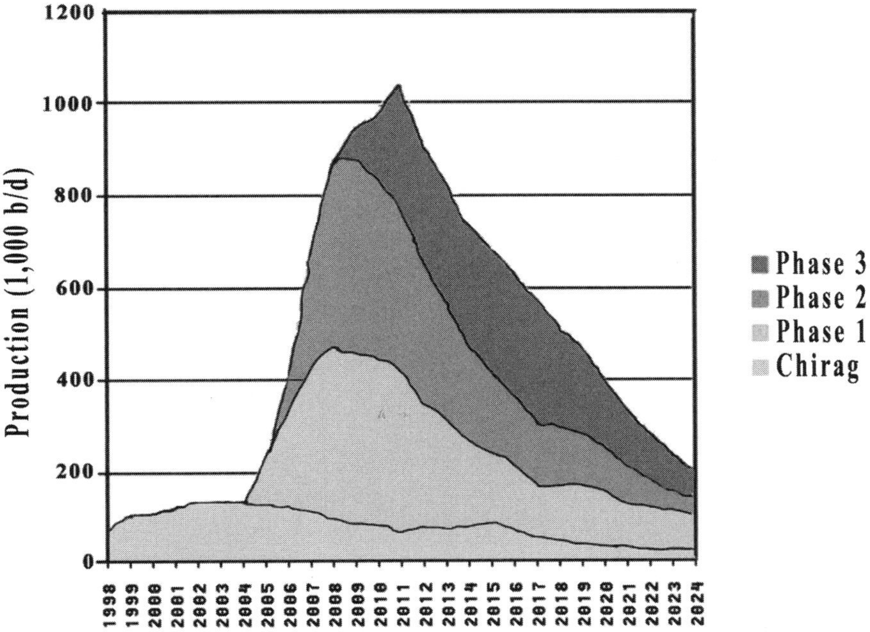

Figure 2.3. Azerbaijan's Projected ACG Production Profile
Source: BP, *Azeri, Chirag & Gunashli Full Field Development Phase I*, February 2002. Reproduced with permission.

In contrast to the ACG fields' brief peak production level, the exploitation of Kashagan calls for a 10-year peak production of 1.2 mmb/d beginning in 2015, as shown in Figure 2.4. The field's production has already been delayed from its original target startup date of 2005. It is now expected to begin producing in 2008, although that date could slip to 2010 or later, given the complexities of the geology and ongoing issues between the consortium members and the Kazakh government. Transportation methods have yet to be addressed, although there is a commitment in principle to move Kashagan volumes (no barrels have been officially committed) through the BTC pipeline after the ACG fields' output begins to decline.

Growth in Caspian domestic consumption will likely be closely linked to the future rates of domestic economic growth. The higher the domestic consumption, the lower is the exportable surplus. Should these economies grow and increase their oil demand to 1 mmb/d or higher as some have suggested,[41] then estimated export volumes derived from the production numbers in Table 2.3 (*infra*) would have to be correspondingly reduced. So far the trend in energy consumption in these countries has been declining rather than rising: in the years of independence for which there are data,

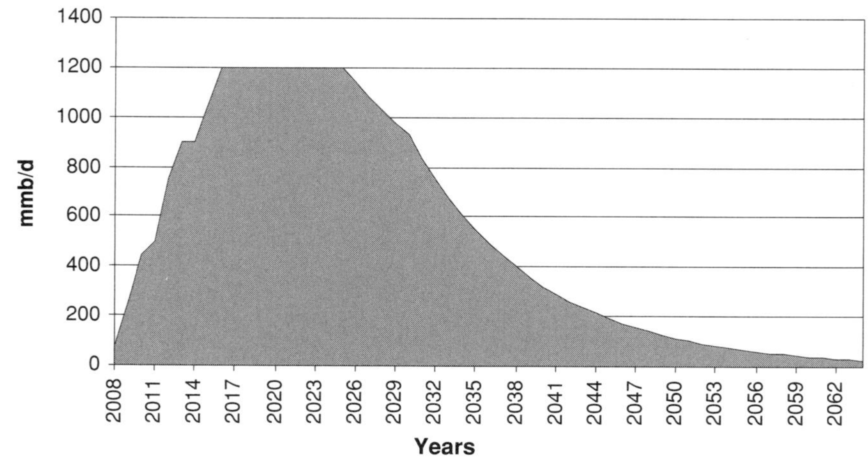

Figure 2.4. Kashagan Oil Field Proposed Production Profile
Source: Nick Griffin and Caroline Cook, *Kashagan—Catch of the Day?* Deutsche Bank, February 25, 2004, charted from Figure 2: 1 (5). Private Report. Reproduced with permission.

the group's internal energy consumption fell sharply from 849,000 b/d in 1992 to 501,000 b/d in 2004, and none of the countries came close to regaining its former Soviet-era consumption level.[42] This author concludes that in the period up to 2020, local regional oil consumption will not likely exceed 500,000–600,000 b/d, and each country will try to monetize its hydrocarbons by exporting as much as possible.[43]

Our production estimates reflect the fact that oil and gas development projects in this region have had—and almost surely will have—a tendency to fall behind schedule. Early optimism on production has had to adjust to the slow and protracted process of negotiations between companies and governments. In addition to the delays in Tengiz, those on various pipelines, and those in Karachaganak and Kashagan, Shah Deniz gas sales and its SCP pipeline were also pushed into the future, as much owing to the absorption capacity of the Turkish gas market as to problems of a political or technical nature.

Russian–Kazakh partners in other offshore and shared fields seem to be in no hurry to start drilling or committing capital, and await acceptable PSAs approved by the Kazakh government. These oil prospects are for the most part low in API degree gravity,[44] which reduces their value. In addition, high metals and sulfur content as well as large quantities of associated gas characterize the prospects, with both the impurities and the associated gas tending to raise costs. Installation of gas treating and gathering infrastructure is likely to be costly as well, according to industry experts.[45] Moreover, the geology of these deep and high-pressure fields is complex and challenging, and ecological concerns are becoming increasingly important. Development of the major fields calls for reinjection of sour gas to maintain pressure. This is a new and untried technique that may or may not work. Finally, a number of oil and gas pipelines run through regions of both high seismic activity and civil unrest, and are at risk of sabotage and disruption. Any interruptions in transport will force reductions in output, affecting production targets and schedules.

ENERGY SECURITY AND FORECASTS OF WORLD OIL CONSUMPTION AND PRODUCTION

Energy security is a concept that has never been precisely defined, and often means different things to different observers. In the broadest sense, energy security can mean a reliable and adequate supply at reasonable prices to support growing populations and economies' performance. While these terms are somewhat subjective and may be interpreted differently by producers and consumers, who are generally not one and the same, the language is typically taken to point to oil and gas prices that are cost-based and determined by market forces. Consumers tend to worry about supply security, while producing countries are concerned for the possibilities of inadequate demand, that is, of demand security. Supply and demand may vary in the short run, but in time relatively high prices signal, for example, the likelihood of reductions in demand and/or increases in supply. In addition to the dimension of time, the state of technology, various transport options, diversification of energy trade, and sources of imports have key roles to play in reducing vulnerability to changes in supply and demand. Sharp rises in prices from inadequate supplies can have destabilizing macroeconomic effects on gross domestic product (GDP), slowing growth in consuming countries and increasing capital inflows, with often undesirable results in producing countries.[46]

We look at the potential production of oil and gas from these Caspian countries, and compare it to forecasts of future world supply and demand. We use the estimates of EIA and the IEA, for they are among the most frequently cited sources for projections of world oil consumption and production. Using the previously projected levels of Caspian oil production from Table 2.2 (*infra*), we show in Table 2.3 the Caspian region contributing approximately 3–3.5 percent of the world's total oil consumption and capacity in the years ahead. This percentage is not likely to alter the central role of the Middle East in meeting future world oil demand.

Any attempt to forecast future oil demand and supply is likely to fail as to accuracy. Yet while the result may be wrong as the future unfolds, it can turn out to be relatively correct, in terms of direction and size. Much depends on expected prices and whether these become reality. What is clear is that by the end of this decade a number of new oil projects all over

Table 2.3. World Oil Consumption and Production, 2010–2020, and Caspian Oil as Percent World Consumption and Production ('000 b/d)

Year	2010	2015	2020
World Consumption—EIA	94.6	103.2	111.0
—IEA	90.4	106.7	121.3
World Production Capacity—EIA	96.5	105.4	113.6
Caspian Oil Production	3.09	3.75	3.36
Caspian as % World Consumption	3.2–3.3%	3.5–3.6%	2.8–3.0%
Caspian as % World Production Capacity	3.2%	3.6%	3.0%

Sources: Percentage estimates by author. World consumption and production capacity estimates from DOE/EIA, *International Energy Outlook 2005*, Table A4, World Oil Consumption by Region, Reference Case, 1990–2025, and Table E1, World Oil Production Capacity by Region and Country, Reference Case, 1990–2025. IEA sources: *World Energy Outlook 2004*, Table 3.1, World Oil Demand. IEA does not provide world production capacity, but gives in Table 3.5, World Oil Supply, the same forecasts as its projected World Oil Demand in Table 3.1.

the world are expected to be on-stream, including developments in Russia, Iran, Iraq, Libya, Angola, Nigeria, Venezuela, and Brazil, among others. There is no long-term world shortage of oil. Future production estimates vary, with EIA projecting a net increase of 11–12 mmb/d in oil production in 2010 over 2002 levels, one private forecaster suggesting in 2005 that the net increment relative to 2004 production capacity was likely to be closer to 16.5 mmb/d in 2010, and other private forecasters suggesting in 2005 that net capacity would be from 13 mmb/d to more than 16 mmb/d greater by 2010 than in 2005.[47] About 1–2 mmb/d of these new amounts is Caspian production, which rises from the 2004 level of about 2.0 mmb/d to 3 mmb/d in 2010, in our view and that of EIA, to as much as 4 mmb/d (including liquids) in others' view. All these estimates of new supplies suggest that supply may far outstrip demand, putting downward pressure on price, and thus dampening the ardor for increases in Caspian production.

Whatever the uncertainties of world production and demand, the Caspian region will remain a minor contributor. From an energy security perspective, Caspian production is a source of diversification of world oil supplies, but it is and will remain only a small player on the world scene. As one international oil executive remarked in private conversation, it is nice to know the Caspian is there, since the region offers an alternative oil source should there be production problems in Venezuela, Nigeria, Angola, parts of the Middle East, or elsewhere in the world.

MONETIZING CASPIAN OIL AND GAS RESOURCES: TRANSPORTATION IS KEY

Based on the conviction that sufficient oil and also gas reserves existed to permit profitable development for sale on international markets, company representatives and many outside observers have over the years offered a multitude of pipeline proposals to monetize these deposits. While Caspian oil is presently moving to Western markets, Caspian natural gas is not, with the exception of Turkmen gas reaching Ukraine through Russia. With the future completion of the SCP, Azeri natural gas will reach Turkish and perhaps other Western markets.

All but two pipelines carrying Caspian crude oil—the Baku–Supsa and the BTC lines—pass through Russia. All but one gas pipeline is controlled by Russia as part of the old Soviet gas pipeline system; the exception is a short line from Turkmenistan to Iran. Nearly from the beginning of the discussion of Caspian export pipeline routes, the United States played an active role in trying to foster routes for Azerbaijani crude that would not cross either Russia or Iran. Yet the United States had no policy objections to the agreement of the Tengiz partners to construct the CPC line from Tengiz across Russia to Novorossiysk on Russia's Black Sea coast. Until CPC was built to carry Tengiz output, field production moved primarily by rail to Black Sea ports.[48]

Pipelines and Risks

Numerous proposals for Caspian oil and gas pipelines have surfaced.[49] Many of the ideas put forth had questionable economics, and some remain bedeviled by concerns over profitability in view of the size of reserves, the uncertain delivered price of oil, and the expected throughput. Investments in field developments or in pipelines in this part of the world are subject to a number of risks. Proper risk assessment is essential to determine whether a given project might be or is economic. These risks include those of geography, geology, pipeline utilization rates and economics, political risks faced by investors at home and in the host country, production costs and variation in oil prices, and the ability of customers to actually pay for their purchases.[50]

- Geography is a major concern to all would-be or actual participants in pipeline projects. Once a pipeline is in place, it cannot be moved, and the countries through which it runs may choose to abrogate or revise the original agreement, exercising monopoly power to increase their take in transit fees.
- Geology always matters, in terms of the size, depth, structure, and cost of reserves. With more exploratory drilling, it has become clear that the

Southern Caspian does not appear to be a promising oil region (the ACG fields date from Soviet times). The northern Caspian region, however, is more promising for both oil and gas.

- Once an oil pipeline is built, its size and capacity—generally measured by diameter and pumping capability—and the speed of filling it affect its profitability. Typically increasing returns to scale exist, in that as the diameter increases (up to a point), throughput costs drop, bringing down the cost per delivered barrel. There is always the risk that other large pipelines could be built that would deprive the first one from realizing its full economies of scale. Any pipeline must be brought into full utilization as quickly as possible from completion to ensure viability.[51] A completed pipeline, such as the Odessa–Brody line, lying empty or only partially utilized, offers insufficient return on the investment outlays incurred, and is unlikely to cover even variable or maintenance costs. Large pipelines suffer from lumpiness—many must be built large or not at all. Others, however, are designed for throughput to increase in stages, should demand warrant, where the stages call for more pump stations; looping the line; or introduction of chemical reagents known as drag reduction agents (DRAs). During the years of discussion prior to the decision to move forward with the BTC line, visitors to the American Embassy in Baku often mentioned posters displayed there that read "Happiness Is Multiple Pipelines." That may be true, in that no producer wants only a single monopolist-controlled route, but it is possible that economic considerations might justify only one appropriately sized pipeline.
- Natural gas pipelines have also been proposed, either new or rehabilitated, for gas exports from the region. These have additional risks, since natural gas is far more costly to move over long distances than is oil. According to one estimate, while it costs roughly $0.50 to move a barrel of oil 1,000 km by pipeline (depending on diameter and number of pumping stations), it would cost from $2.50 to $4.00 to move the same amount of energy in the form of natural gas over a similar distance.[52] In addition, long-distance gas pipelines need a proven market at the destination, and in this sense also differ from oil pipelines, where at the end of the pipe oil can be loaded onto a tanker to reach world markets. In the Caspian as elsewhere large gas fields exist whose reserves are, or may be, "stranded" and remain unexploited, in that the gas cannot be developed and transported at a price high enough to permit production yet low enough to win customers.
- Political risk permeates Caspian energy development projects. In general, political risk can be divided into two types: first, what foreign companies face in the host countries; and second, what they face from their own home governments. Political risk is a great concern in the Caspian region, where laws and institutions are weak; governments challenge contracts that companies considered ironclad; fiscal regimes change

unpredictably; environmental fines are imposed somewhat whimsically; ethnic, clan, and various civil conflicts occur; corruption is rampant; and regimes are ever more authoritarian. To offset the risk that is itself difficult to quantify, the project's rate of return must be higher than normal; this risk premium alone has been estimated at around 10 percent.[53]

For American companies operating overseas, the second type of political risk arises from pressure from a U.S. administration or from the U.S. Congress. This may take the form of pushing companies to pursue pipeline projects that may have strategic, but no commercial, appeal. In addition, the United States can prohibit its national companies from engaging in business activity with or in a given country. U.S. sanctions against Iran, for example, effectively rule out U.S. company participation in proposed export pipelines from the Caspian to the Persian Gulf, even though the Iranian route was and is widely believed to be the cheapest of all the oil export alternatives proposed. One of the hopes of U.S. international companies active in the Caspian is that some day an improvement of United States–Iran relations will permit reentry into the Iranian oil and gas market. Such a future, of course, depends in key part on what steps the new government of Iran takes to attract and keep foreign investment in the oil and gas sector, as well as on U.S. foreign policy decisions. Since Iranian reserves are thought to be far less costly to develop and transport than their Caspian counterparts, companies interested in Iran may yet find themselves with a high-cost Caspian white elephant on their hands.

- Expected oil and gas production costs and market prices are generally not known with certainty, and thereby pose risks to producers and transporters. Both oil and gas are normally sold at prices that must be competitive in the customers' markets, which may have other sources of supply. This competition generally results in netback pricing, with the wellhead price being the residual received after transport and other costs have been deducted from the delivered price. If the costs of getting oil or gas out of the ground are relatively low, the transport cost could be somewhat greater than average, with the delivered price still remaining competitive. As costs of production rise and market prices do not, the pipeline's profit margins decline. If market prices fall, and transport prices do not, the producers find margins squeezed and perhaps risk inability to recover costs, or an inability to generate the target rate of return. These risks become sizeable if a pipeline is under consideration for transporting oil or gas that is known (or expected) to be high-cost to extract. Price volatility in oil and gas markets is substantial and common. When oil prices fell to near $10 per barrel at the end of 1998, the economic viability of a number of pipeline plans and exploration projects was challenged. For planning purposes, some writers have estimated that the combined production and transport costs for Caspian oil should not exceed

$12–15 per barrel, while others say $15–18 per barrel, and still others say costs and transport are as high as $20 per barrel in real terms.[54]

- For markets to work, those who agree to buy must pay, and those who agree to sell must deliver—assuming all other contract terms are met. Simple as these principles appear, they are often not adhered to in the Caspian region. Gas customers like Ukraine have been known to fall behind in payments, or to fail to pay at all, acting as if the gas were a free good as it was in Soviet days. In most international gas markets, the buyer signs a long-term take-or-pay contract with the producer, who uses this guarantee of revenues as collateral to secure financing to build the pipeline.[55] But customers in this region often do not have the credit worthiness to enter into a take-or-pay contract. Turkey's overestimated economic growth has already caused it difficulties in meeting payments for gas delivered under its various take-or-pay contracts. Turkmenistan is rich in gas, but may have committed future sales volumes to Russia and Ukraine that are larger than its known reserves; it may yet walk away from its signed contracts, leaving pipelines underutilized and customers undersupplied.

In summary, oil and gas reserves are nearly worthless if transportation cannot be arranged. International oil companies and governments alike face major project challenges in identifying markets, routes, costs, and financing, and ensuring that production and transport capabilities are completed simultaneously.

Transport Proposals

Ever since the demise of the Soviet Union and the growth of interest in hydrocarbons in this area, there has been a flood of pipeline and other transport proposals. We divide transportation proposals into four categories: those that have been built or are under construction, those that might be constructed or rehabilitated over the next 10 years or so, those that are unlikely to be built in that period, and those unlikely to be built at all.

1. Completed or under-construction oil pipelines include the following:

- The 830-km line from Baku, Azerbaijan, to Supsa, Georgia. It has operated since early 1999, and is called the western route for "early oil" exports.[56] This line was a preexisting former Soviet Union oil pipeline in disrepair. AIOC eventually spent roughly $600 million to rebuild and improve it. The line carries about 150,000 b/d. From oil storage tanks at Supsa, tankers load for shipment across the Black Sea and through the Turkish Straits to Western markets. This line was one of the competing alternatives to BTC, with proposals which called for expansion

Figure 2.5. Selected Caspian Crude Oil Export Pipelines to the Black Sea and the Mediterranean
Source: U.S. Government 764917AI (C00733) November 2003.

to at least 500,000 b/d when Azeri production was forecast to mount sharply.[57] Figure 2.5 depicts the route of this line, along with other lines carrying Caspian crude oil westward.

- The northern export route from Baku (also shown in Figure 2.5). This line runs for about 1,400 km northward through Dagestan and a bypass of Chechnya (picking up some barged oil arriving at Makhachkala) before bending northwest to Tikoretsk, and then southwest to Novorossiysk. Also a former Soviet oil pipeline, it was a target of Chechen dissidents. Despite the Chechnya bypass of over 280 km, built in 1999–2000, throughput amounts to only about 50,000 b/d, which is the volume that SOCAR ships. Were actual capacity equal to nominal capacity of 180,000 b/d, shippers would still prefer not to use this line, not only because of political strife in the region, but also because Russia offers no quality bank to compensate for the mixing of light Azeri oil with heavier and less valuable Russian crudes. Over the years, the Russians have proposed extensive upgrades and expansion to this line to bring it to a 1-mmb/d capacity—in effect, a complete rebuilding. Had that expansion actually occurred, the line could have served as the main export pipeline for ACG production, had quality-bank issues been resolved. At Novorossiysk, oil moves by tanker through the Turkish Straits. With the opening of the BTC pipeline, the northern route is unlikely to be used regularly, but kept as backup capacity. Flows can continue in bad weather, since the terminus has storage capacity to overcome the port of Novorossiysk's closure for as much as 2 months a year due to pack ice, which prevents tanker loadings.

- The BTC pipeline (also shown in Figure 2.5). It runs some 1,768 km from Baku, Azerbaijan, northwest to Tbilisi, Georgia, then southwest (avoiding Armenia) to Erzurum and Ceyhan, a port open all year. Completed in May 2005 at an estimated cost of $2.9 billion (associated tankage and terminal costs lift the total to about $3.6–4 billion), first tanker shipments from Turkey were expected to begin in the second half of 2005 but slipped until the first half of 2006. Over time, the completion date was extended, with construction delayed by both weather considerations and environmental challenges. The line has an initial capacity of about 1.2 mmb/d, with expansion if merited to nearly 2 mmb/d with looping, additional pump stations, and DRAs.[58] AIOC, led by BP, chose the project over other alternatives in 2002 as the main export pipeline for ACG output. For the better part of the 3 years prior to the BTC route selection, AIOC preferred the Baku–Supsa and/or Baku–Novorossiysk options, both of which were far less costly to improve and expand than to construct the new BTC line.[59] Both of the existing routes, however, required further transit through the Bosporus. The cheapest alternative route to the proposed BTC was one to Iran's Kharg Island export terminal, but that was ruled out due to U.S. policy toward Iran and the participation of U.S. companies in AIOC. In the face of unremitting pressure from the U.S. government, and Turkey's desire for the employment and transit fee benefits of a line through its territory that did not require Bosporus passage, AIOC finally changed its position and indicated that it would pursue construction of the BTC line as a "strategic" endeavor.[60] Some observers believe that the risk of a spill or accident in the Bosporus, potentially permanently damaging BP's reputation, was the telling argument that led to the selection of BTC. This project did not please Russia, which wanted to earn transit revenues through its territory. Moscow finally dropped its objections to BTC when AIOC stated that it expected the line to be economic.

The debate both before and after the decision to build BTC, however, focused on whether there was enough oil in the Azeri ACG fields to support profitable operation of the line. New oil discoveries in Azeri waters had not occurred. It is of interest that not all members of AIOC chose to belong to the BTC consortium, while non-AIOC participants joined it.[61] Those members of AIOC that chose not to participate in BTC were ExxonMobil and Devon, while Total, ENI, and ConocoPhillips—not members of AIOC—joined BTC. ExxonMobil announced in November 2004 that it would not ship on BTC, but would move up to 200,000 b/d by rail through Georgia to the Black Sea.

The ACG production profile (Figure 2.3, *infra*), however, shows a rapid rise to peak output levels of just over 1 mmb/d, followed by nearly as sharp a decline in the years after 2010. This means capacity will be available on BTC, and it is hoped that producers in Kazakhstan will fill it. No formal commitments of specific volumes have been made, but press reports indicate plans to barge as much as 200,000–400,000 b/d to join BTC initially, and at some future date to move oil by undersea pipeline.[62] How, and how much, Kazakh oil moves to Baku will depend on other transport alternatives, including costs, timing, and size. The shipment of Kazakh oil via BTC is one of the most frequent "on-again, off-again" deals, and at the moment it is "off."

The BTC project continues to face political and environmental challenges as well as economic ones. Given the ongoing instability and risk of war in Georgia with its secessionist provinces, and the Nagorno-Karabakh conflict in Azerbaijan, the BTC pipeline and its companion SCP gas pipeline may be at future risk of terrorist attack, or of sabotage and interruption.

- The under-construction SCP pipeline of 690 km. This gas line calls for delivery of 6.6 bcm/y of Azeri Shah Deniz gas to Turkey, beginning in late 2006 (completion is now 2 years later than the initially proposed date). The project permits expansion to carry at least 16 bcm per year, and some sources suggest that given the field's large reserves the line could carry as much as 30 bcm annually.[63] The line will follow the BTC route as far as Erzurum in Turkey, where it will join the Turkish gas network. While there is a consensus that Turkey remains overcommitted to gas from other sources as well as from Shah Deniz, the project is going forward.[64]
- Atyrau, Kazakhstan to Samara, Russia, a preexisting 720-km Soviet oil line. Under consideration for expansion, it is used to carry Kazakh oil northward to join the Russian pipeline and export system. Currently its capacity is about 320,000 b/d, and expansion to 500,000 b/d is the goal if the feasibility study is positive, producing companies guarantee supplies, and space exists on connecting Russian pipelines. The Samara line is a key link for northern Caspian producers, as shown in Figure 2.6.

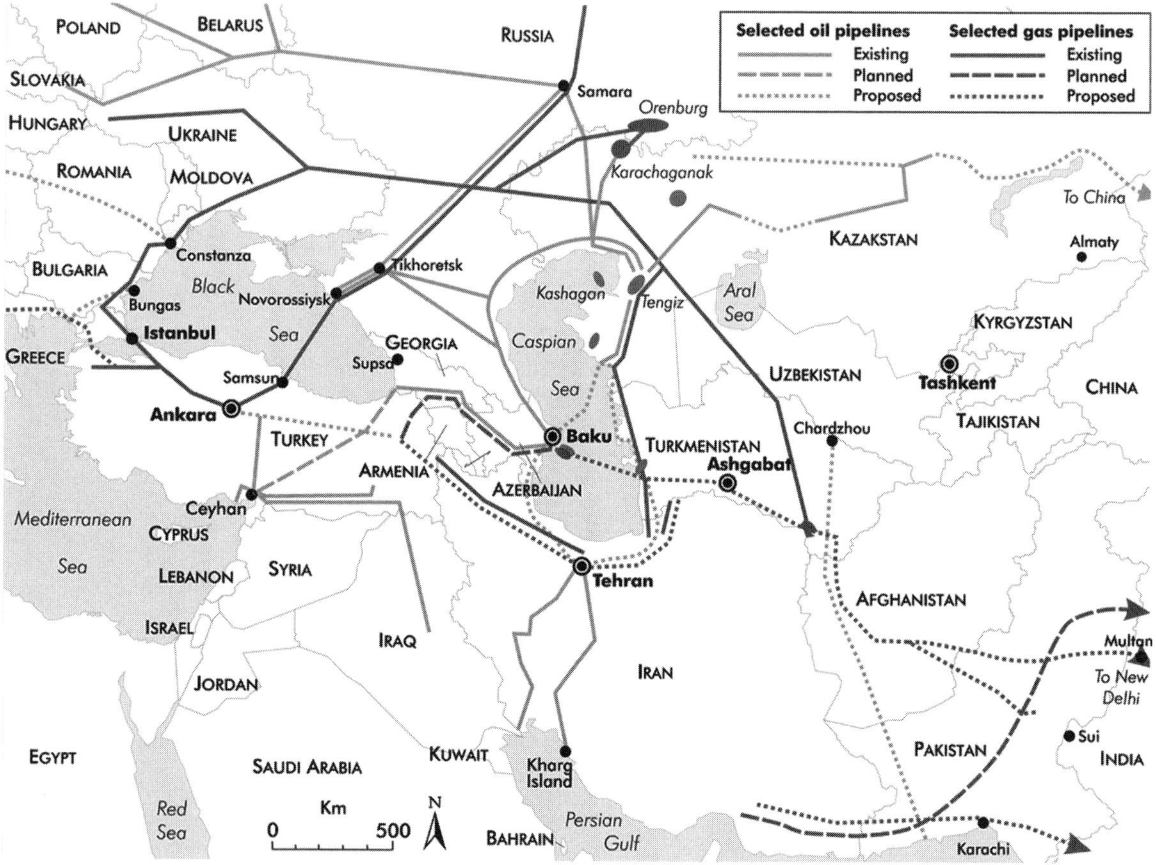

Figure 2.6. Caspian Region Oil and Gas Pipeline Export Routes

Source: International Energy Agency, *World Energy Investment Outlook, 2003 Insights*, 154. Reproduced with permission.

Figure 2.7. Pipelines Serving Tengiz and Karachaganak Fields
Source: ChevronTexaco Website. Reproduced with permission.

- The 1,580-km CPC oil pipeline carrying crude from the Tengiz field in northeastern Kazakhstan to Novorossiysk, Russia.[65] The only privately owned pipeline that runs through Russia, it opened in late 2001. The line's initial capacity was 565,000 b/d, with a design capacity of 1.3–1.4 mmb/d. By end-2003 the line was operating at nearly full initial capacity. As of the first half of 2005, the shareholders decided to expand the line to full design capacity, although differences with Russia caused a delay in reaching that decision. With additional pump stations and the introduction of DRAs, capacity could probably be increased to 1.9–2.0 mmb/d. CPC members also built an offshore tanker-loading facility northwest of Novorossiysk. Figure 2.7 shows the routes and interconnections of the various export lines.
- The approximately 650-km oil pipeline from the Karachaganak field in Kazakhstan. It joins the CPC line at Atyrau, Kazakhstan. Completed in 2003, its capacity is 225,000 b/d. The field produces both oil and gas, with oil seeking Western markets, and up to 7 bcm/y of gas sold to

Russia.[66] Following the start of production in 2003, Karachaganak oil moved over both the CPC and Atyrau–Samara lines.

- The 674-km oil pipeline connecting Odessa on the Black Sea in Ukraine to Brody, in northwestern Ukraine, and potentially to be extended northwest to Plock and Gdansk, Poland, and/or to Plock and Wilhelmshaven, Germany. The issue for some time has centered on whether Caspian oil will flow north from Odessa, or Russian oil will flow south from Brody and onward through the Turkish Straits. As of the time of writing, it appears that the current flow of Russian oil south to Odessa will be terminated by 2008, the pipeline reversed, and Caspian oil could ultimately flow in the originally proposed direction. If issues of the quality bank are not resolved, however, and Caspian oil is mixed with lower-quality Russian oil at Brody, Caspian deliveries will likely lose up to $5 per barrel of their value, and thus could discourage shipment from Odessa. This directional flow has major consequences for increased volumes passing through the Bosporus.

For the northern alternatives, both Gdansk and Wilhelmshaven offer deep-water ports. Wilhelmshaven to the west has the environmental advantage over Gdansk since the former destination does not require shipping around the Danish islands and peninsula. Deliveries to Wilhelmshaven would pose a direct competitive threat to North Sea oil. The Odessa–Brody initial capacity of 180,000 b/d can be expanded to 500,000 b/d and again to 900,000 b/d, should the line be built to Plock, where there is a refinery and connections with existing crude oil lines serving northern Europe. If this system is utilized, it will relieve pressure on transit through the congested Bosporus, and in that sense can be considered one of the proposed Bosporus bypasses. The Ukrainian government built the line in 2001, but it lay idle for about 3 years, initially for lack of Caspian oil commitments and the unsettled issue of which company or government was to pay for the line fill. Kazakhstan was thought to be ready to ship 120,000 b/d on Odessa–Brody, and Azerbaijan had also expressed an interest in using the line, but neither has formally agreed.[67]

The directional issue was purportedly resolved in February 2004, with oil to flow north from Odessa. Ukraine and the EBRD were thought to be close to reaching agreement on financing the extension to Gdansk.[68] But only 6 months later, Ukraine decided to use the Odessa–Brody pipeline in the southerly direction, thereby providing an additional export outlet for growing Russian oil production. With the decision to reverse the flow, Caspian export plans came to naught. The change in plans was a direct blow to Washington and Ankara, both of whom supported the northern flow as an alternative to increasing volumes transiting the Turkish Straits. But in February 2005, the new Ukrainian government revived the idea that the pipeline should carry Caspian oil in the original direction, and began working with Poland to extend the line eventually to

Plock and to Gdansk, for use after 2008.[69] The issue, however, remains in flux, as no firm plans for direction or northern extension have been announced.

- Pipelines receiving swapped oil. Iran is presently receiving Caspian and Russian oil by barge for swap at its northern port of Neka in exchange for the equal value of barrels sold at the Persian Gulf. Caspian supplies are now refined in northern Iran,[70] and to date have not been deemed to violate U.S. sanctions. Iran uses the $1.50–2.00 per-barrel swap fee to fund upgrades necessary for processing these high-sulfur crudes in its refineries at Tehran and Tabriz. Tehran says that it can accommodate swaps of up to 370,000 b/d of Caspian crude and has plans to raise swap capacity to 700,000 b/d–1 mmb/d. Current swap volumes, however, are far lower, at approximately 150,000–170,000 b/d, from Russia, Kazakhstan, and Turkmenistan.[71] Swap volumes are expected to grow, particularly with the construction of new Caspian shuttle tankers, as long as robust Russian production continues to need outlets beyond the capacity of the country's pipelines and ports. Press reports in mid-2005 suggested that Iran offered unnamed Russian companies 25-year swap arrangements for up to 300,000 b/d, although contracts were not signed.[72] In 2004, tankers experienced waits to unload at Neka, so brisk had the trade become. With swap costs far lower than BTC tariffs, Iran is mounting a major competitive alternative to BTC, even though some shippers may prefer pipeline shipments to the west rather than swaps to the Persian Gulf.[73]

2. *A second category of pipelines includes those that have a chance of being built between now and 2015, although observers differ in their estimates of the probabilities.*

This category includes the various and competing proposals for a Bosporus bypass (other than Odessa–Brody), one of which could be built in this time period, a new and dedicated oil export pipeline for Kashagan production if needed, and proposed oil and gas export pipelines to China, which will require rehabilitation and extensions of existing Central Asian pipelines. Turkey is concerned that increasing tanker traffic through the Straits will raise the risks of accidents and pollution.[74] In view of these concerns, some observers doubted that the CPC line delivering Kazakh oil to Novorossiysk would be expanded to design capacity unless new pipelines avoiding the Bosporus are built. The decision to expand to 1.3–1.4 mmb/d means bypasses will almost certainly be needed. Bypasses could also accommodate the mounting volumes of Russian crude departing Novorossiysk.

Ideas for Bosporus bypass pipelines abound, and most have been around for years. The suggested routes all terminate at ocean ports of varying capacities.[75] In recent years about 3 mmb/d of crude and products passed through the Bosporus, and some observers forecast a level of closer to 4.0–4.5 mmb/d by 2010, in the unlikely event that the Turks were to permit such volumes. The Turks have, of course, a financial interest in terms of transit fees, not only in BTC, but also in any future bypass pipelines crossing Turkish territory. At present, there is no fixed capacity limit to the Straits; to all intents and purposes it is what the Turks can be (politically) made to say it is, and this will depend on their regulations governing length, size and spacing of ships, tug escorts, Turkish pilots if required, refusal to permit nighttime passage, and other stipulations they succeed in imposing under the Treaty of Montreux. Both the United States and Russia, however, would oppose extreme capacity limits; the two of them have the power to exert severe pressure on Ankara. Nonetheless, the congestion delays of up to 3 weeks in the winter of 2004, demurrage charges, and concerns over potential new regulations for tanker passage generated some rethinking on whether and when a bypass pipeline makes economic sense. Future tanker traffic and delays are likely to increase in the absence of a bypass line. There is, however, a free-rider problem: Why should a shipper incur an additional bypass tariff of about $1 per barrel so as to permit competitors to use the now less-congested Turkish Straits for free?

When externality costs become too great for Bosporus and Dardanelles tanker passage, a bypass is likely. None of the bypass alternatives, however, is itself without environmental risk, since spills can occur as oil is transferred between ships and pipelines. Despite its anticipated revenues from bypasses, Turkey argues that the externality costs of Bosporus passage are already too great and that a bypass is necessary as soon as possible. Ankara, however, does not want to repeat the experience of Ukraine's Odessa–Brody line with any of the bypass lines idled once one or more are constructed. Moreover, an additional consideration for Turkey is that as long as there is idle capacity on the BTC line, Turkey may not take any action that might be seen to undermine that project; Ankara's endorsement of a bypass route might be viewed as a violation of the BTC agreement.[76] Nonetheless, it now appears that a bypass route is under serious consideration, and will eventually be built.

Figure 2.8 shows some of the various bypass alternatives, one of which may be built within the next few years. The first three bypass proposals call for new lines lying fully within Turkey, simplifying the timetable since international agreements with other countries are not needed, and all permit further onward transport in 300,000 dead-weight-ton tankers. Despite the 2005 fanfare accompanying the intergovernmental agreement on the Bourgas–Alexandroupolis bypass, it is far from certain that it is the front-runner or that the line will be built. Whether any of these proposed bypasses will be realized, however, also depends on alternative routes and capacity, such as the oil pipeline from Kazakhstan to China, initial use and proposed extension of the Odessa–Brody line to Gdansk and/or

Bypass Proposal	Host Countries	Planned Capacity
Burgas - Alexandroupolis	Bulgaria, Greece	300,000 to 700,000 b/d
Burgas - Vlore	Bulgaria, F.Y.R.O.M.*, Albania	750,000 b/d
Constanta - Trieste	Romania, Serbia, Croatia, Italy	700,000 b/d
Kiyikoy - Saros	Turkey	1 to 2 million b/d
Odesa - Brody - Gdansk	Ukraine, Poland	240,000 to 800,000 b/d
Samsun - Ceyhan	Turkey	500,000 to 1 million b/d

Figure 2.8. Proposed Bosporus Bypass Pipelines
Source: U.S. Government 766111AI (R02737)/392839 March 2004.

Wilhelmshaven, the expansion to 1.2–1.3 mmb/d at the Primorsk, Russia, crude oil export terminal, as well as local Turkish and Greek environmental opposition, citing risks to tourism. These proposals include the following:

- Kiyikoy on the Black Sea to Ibrikbaba on the Aegean Sea, known as the Trans-Thrace pipeline. This route is about 200 km long. The proposal calls for a capacity of 1.0–1.2 mmb/d and was initially supported by Russia's Transneft oil pipeline monopoly. TNK-BP allegedly agreed to guarantee oil for the $900-million line.[77] Despite Turkey's interest in reducing oil tanker congestion in the Turkish Straits, Ankara has been unwilling to commit funds for this line. The route is nearly parallel to the one described below. Transneft had originally planned on participating in construction, but backed off, citing lack of progress with the Turkish authorities.
- Igneada on the Black Sea to Saros, Turkey on the Aegean Sea, with a proposed capacity of 1.0–1.5 mmb/d. The Thrace Development Co. expects to win a license from Ankara to build the 200-km line. This route is supported by Kazakhstan and the United States, and has the backing of businessmen in Turkey and the United States. Tariffs would be about $1 per barrel initially, falling to half that level after recovery of capital costs. Turkey has announced its intention to build both the Igneada–Saros line and the line from the Black Sea port of Samsun in Asian Turkey to Ceyhan on the Mediterranean, although scheduling and financing have yet to be determined.[78]
- Samsun to Ceyhan, Turkey, via Ankara (about 860 km), with a proposed capacity of 500,000 b/d to 1 mmb/d. This route also has the backing of the United States, and permits tanker loadings from Ceyhan to world markets. If this line is built, Russian and Caspian oil received at Ceyhan could reach Asia by a route shorter than around Africa, were the 400,000 b/d TIPline from Eilat on the Red Sea to Ashkelon on the Mediterranean to be reversed.[79]
- Bourgas, Bulgaria, to Alexandroupolis, Greece (285 km), with an initial capacity of 700,000 b/d, expandable to 1.0 mmb/d. The line is estimated to cost $1 billion. Russia, Greece, and Bulgaria signed an intergovernmental agreement in April of 2005, although the signing does not necessarily guarantee the line's construction.[80] The project is to be financed privately. Doubts about the project's economic viability remain, and financing will depend on whether enough volumes are committed to the line. Greece allegedly stands ready to contribute roughly $130 million to the line's construction, which could start within a year and be completed in 3–4 years.
- Bourgas, Bulgaria, to Vlore, Albania (about 900 km), with a proposed capacity of 750,000 b/d, and expected to cost at least $1.1 billion.

- Constanta, Romania, to either Trieste, Italy, or to Omisalj, Croatia (about 1,375 km), with a projected capacity of 700,000 b/d–1 mmb/d. This line is viewed as a competitor to Odessa–Brody in ultimately serving European markets.

In addition to Bosporus bypasses, other pipelines have been proposed, some of which are likely to be constructed. These include primarily a new pipeline to handle output from the Kashagan field in Kazakhstan, with a peak production level of 1.2 mmb/d expected in 2015–2016, if it stays on schedule; a trans–Caspian Sea line from Kazakhstan to Baku; a new crude export line from western Kazakhstan to China; and new natural gas export routes. We address these proposals in turn.

- Will production from Kashagan require a new export line and, if so, when and where? Some would say that there is sufficient expandability in existing lines that no new pipeline system will be required. As the respected Scottish consultancy, Wood Mackenzie, argued in 2004, there is adequate capacity in the BTC, CPC, Odessa–Brody, Samara, the northern and western routes from Baku, and/or barge from the port of Aktau in Kazakhstan to the Azeri port of Makhachkala on the northern route, to accommodate Kashagan production, assuming there is a quality bank at Russian termini to preserve the value of light Kazakh crude.[81] While Wood Mackenzie concluded that an additional line for Kashagan output is not needed, it nonetheless noted that new lines could be built provided they could compete economically with existing lines. Others suggest, if they are optimistic, that additional fields will be found near Kashagan and that a new line will be needed, most likely through Iran, regardless of the U.S. political posture toward Iran. Given the uncertainties affecting Kashagan's development profile, we believe the earliest a new pipeline could be completed would be after 2015. Our view is that a new pipeline dedicated solely to Kashagan oil exports will not be needed.
- An oil pipeline across the Caspian Sea to link Kazakh oil production in the north, and perhaps Russian as well, to the BTC pipeline in the south. At present this is referred to by the acronym ABTC, for Aktau-BTC. Were a pipeline constructed, it would roughly parallel the tanker route between Aktau and Baku. The as-yet insufficient volumes, together with the absence of agreed seabed delimitation, undefined costs, and environmental challenges from an unconsolidated seabed, earthquakes, and mud volcanoes have put this proposal on the back burner for now. When rail or barged volumes begin to approach or exceed 400,000 b/d, however, industry experts agree that a pipeline offers lower costs than barges. While there appears to be increasing interest by both Kazakhstan and Azerbaijan in pursuing this project, no volumes have been committed.[82] On the occasion of the BTC pipeline's inauguration, the press reported

that Kazakhstan is prepared to supply "up to" 600,000 b/d to BTC, but timing and modes were not indicated.[83] A pipeline project is likely to go forward but closer to 2015 rather than sooner, since capacity is unlikely to be available on BTC until after its peak in 2010–2012 (see Figure 2.3, *infra*).

- China is already a buyer of Central Asian oil, historically delivered by rail. Its actual and expected rapid growth in energy demand have revived its interest in long-distance oil and gas by long-distance pipelines, from Russia as well as from Central Asia. In 2005 China completed construction on an approximately 1,000-km oil pipeline from Atasu in central Kazakhstan to Alashankou on the Chinese border. It is expected to be fully operational by mid-2006.[84] The proposal for an oil pipeline to China running the width of Kazakhstan would have appeared in the third category in this author's view, were China's oil demand not so strong, and had not the idea of a Russian oil pipeline from Angarsk to Daqing been at least temporarily eclipsed by a proposal for a larger one to Nakhodka, a Russian port that could serve international buyers. Capacity estimates for the Kazakh line to China range from 200,000 to 400,000 b/d, even though a similar proposal in 1997 for a 400,000 b/d oil pipeline was abandoned due to insufficient reserves, high costs, and low oil prices. The cost at that time was estimated at $2.5–3 billion, to be paid for by China, and completed in 2005. In our view, a 400,000-b/d line is too small—then and now—to make economic sense.

Sensible or not in economics terms, some segments are already built. The first 448-km Kazakh section linking Atyrau, which sits at the junction of the CPC and Samara lines, to Kenkiyak to the east, opened in 2003; oil flows currently moving west will be reversed in the future. The section linking Atasu to Alashankou may be extended 252 km to serve the refinery over the border at Karamay in China.[85] The third 2,000-km section calls for linking Kenkiyak and Atasu through Kumkol (an oil-producing region in central Kazakhstan), but the link between Kumkol and Atasu is not yet scheduled for construction. The Kumkol–Kenkiyak portion is nearing completion. China is expected to finance the lines, but details on financing are cloudy. These connections could yet be either deferred or abandoned.

In addition, Central Asian countries hope to win gas exports to China, but pipelines from Turkmenistan, Kazakhstan, and Uzbekistan to and through Russia need substantial repair. The capacities of these natural gas lines are at or below 50 percent of their design levels.[86] Accidental explosions occur frequently. Gas pipeline rehabilitation has not attracted Western investor interest, since the gas does not sell at world price levels and payment is often partially in kind. The shipping countries, together with Gazprom, are beginning to refurbish the lines, but original capacity levels have yet to be reached.[87] Turkmenistan has agreed to sell gas to Russia and Ukraine in such large volumes that questions arise as to whether and how these can be transported, and whether Turkmenistan's gas reserves are sufficient to support the future sales. China's interest in buying gas from Russia's Kovytka gas field—if Russia permits timely Kovytka development—may once again squeeze Turkmenistan's hopes of selling gas to China. Nonetheless, China's 4,200-km West-to-East gas pipeline, which originates at the Lunnan gas fields of the western Xinjiang province (near the border with Kazakhstan), was completed in late 2004 and could be linked to a gas pipeline from Kazakhstan and/or from Turkmenistan.

Finally, we look at the possibility for Caspian natural gas to move west to Europe, through and beyond Turkey.[88] The Nabucco gas pipeline project was initiated by the European Parliament and the European Council in June 2003. It is to run through five countries, with BOTAS of Turkey, Bulgargaz of Bulgaria, Transgaz of Romania, MOL of Hungary, and OMV of Austria each holding 20 percent ownership. It is not certain where either the gas or the line will originate, but the line is to terminate at Baumgarten, Austria, which is a major nexus for existing gas pipelines serving EU countries. OMV's feasibility study suggests that a 25–30-bcm/y project is economic, despite the length of the route (some 3,300 km of new pipe). The estimated $5.6 billion project was originally intended to be finished in 2009, but the timetable has slipped. The final investment decision will not be made until at least 2007, and opening is not expected until 2011. Peak throughput volumes would occur by 2020. The line could be a big boon to the Shah Deniz field, which is estimated to be able to produce 15–30 bcm/y. Iran, Kazakhstan, and Turkmenistan might also sell gas to Europe were this pipeline to come to fruition. The challenges to successful completion are great, for five countries must agree on all aspects of the line and its tariffs, costs are likely to rise, and purchase and supply agreements have yet to be determined. While this project could be operational between now and 2015, we put the chances at less than 50 percent.

3. In the third category are pipelines that are not likely to move forward in the next 10 years or so, but could do so beyond 2015, if production profiles and demand conditions warrant:

- A new main oil export pipeline traversing the Caspian Sea and running south through Iran. One proposal calls for a 1.0–1.5 mmb/d line to carry Caspian oil to Iran's Kharg Island export terminal in the Persian Gulf. Widely recognized as the cheapest on-land alternative for Caspian production to reach world markets, it is unlikely to be built unless and until U.S. relations with Iran improve, in our view.
- A second and parallel BTC line. Once the original BTC is operating successfully, it may be simpler and far less costly to expand existing facilities than to add another line. In our view, construction of a new second line is

not likely, although expansion in the existing line to a capacity of about 1.8 mmb/d is, provided Kazakhstan commits large volumes.

- Trans-Caspian—Turkmenistan to Baku—twin gas and oil subsea pipelines. This project was proposed some years ago, but was abandoned in the face of Turkmen intransigence; the decision to build the SCP line to serve Turkish gas markets; and the recognition that with Azerbaijan's sale of Shah Deniz gas to Turkey, the Turkish gas market would be oversupplied for some time to come.

4. A final category is pipelines that are unlikely to be built:

- Construction of a second pipeline parallel to the CPC line. We rule this out for reasons of overdependence on Russia as a transit country, and of vulnerability to Turkish limitations on tanker passage through the Bosporus.
- A link from Novorossiysk, Russia, through the rebellious Abkhazia province of Georgia to join the BTC line. We do not expect this link to be constructed due to Russian–Georgian politics and Abkhazian instability.
- An approximately 1,680-km, $3.5-billion gas pipeline from Turkmenistan through Afghanistan to Pakistan and possibly to India. Unocal was one of the original project sponsors but withdrew in 1998 owing to political uncertainty in Afghanistan under the Taliban. The ADB in 2005 found the project feasible, if financing could be arranged, although doubts remain as to whether Turkmenistan has sufficient gas to meet Ukrainian and Russian, as well as Pakistani and/or Indian, demand.
- An oil pipeline from Kazakhstan through Uzbekistan and Afghanistan to Pakistan. This proposal suffers from distance, difficult terrain, political instability, and difficulties in arranging financing.

COSTS AND PRICES

Cost information on the various energy development efforts in Azerbaijan and Kazakhstan is proprietary and difficult to obtain. Some estimates drawn from company data have been published, nonetheless, indicating that fully built up costs for the newer offshore areas fall in the $12–15 or $15–18 per-barrel range, well above comparable costs in the Middle East.[89] The consensus appears to be that $20 per barrel in real terms is necessary for profitability. Built-up costs include all the costs of development, transportation, and operation. Costs should decline once capital expenditures are recovered, and when interest charges are no longer included. In some onshore Kazakhstan areas, production costs are far lower, but transportation charges to markets add significantly to the total.[90] In general, finding costs are very low at $0.30–0.50 per barrel, with total costs dominated by transportation and development costs.[91] Typically, transportation charges amount to about half of per-barrel development costs, and in some instances have amounted to as much as $14 or more per barrel for rail transport.[92]

In a period of robust oil prices of $40–70 per barrel, this level of cost compared to price looks extremely attractive, although the opposite was true in 1998–1999 when prices fell as low as $10 per barrel. If one assumes that the government's typical profit share is 80 percent, with a 20 percent share for the investors, then at a price of $30 and a cost of $15, the investors are left with $3 per barrel as their return. Alternatively, at a price of $20, and the same profit-share split, company profits are $1 per barrel, which probably does not justify the investment.[93] Most observers conclude that a price of at least $20 per barrel in real terms is needed to justify overall Caspian investment; should prices fall below this range, new development and production activity is likely to halt, and could not compete with output from the Middle East.

Flow rates and well productivity, however, may be as important as world prices and government take in estimating costs and returns. High flow rates reduce unit costs. Flow rates in the Caspian region have been as great as 5,000–10,000 barrels per well, with one well setting a record of 18,000 b/d in 2002. These rates are comparable to some from the most prolific wells in the Middle East.[94] According to ENI, the operator of Kashagan, the productivity per well in that field is expected to be between 10,000 and 30,000 per day.[95] This geologic advantage, together with technology likely to drive costs down even further, indicates that Caspian oil, at least from the more prolific deposits, can be profitable at approximately $20 per barrel in real terms.

Further evidence on costs, based on conversations with company representatives, suggests that:

- In Azerbaijan's offshore fields, production from ACG remains profitable at a per-barrel wellhead price of $12, but not below that level. This cost may yet be too low. While development and lifting costs per barrel are not public, they are reportedly above expectations and rising fast due to drilling challenges and malfunctions resulting from high salinity, reservoir pressure, and the unstable seabed. While the problems have been overcome, the solution required the use of more expensive technologies than originally anticipated.[96] The $12 per-barrel figure includes a transport fee of about $3.30 per barrel to Ceyhan, and perhaps another $1.50 per barrel for terminal, port, and tanker charges to Mediterranean markets. Pipeline tariffs from Baku to Supsa are estimated at $0.43 per barrel, and those from Baku to Novorossiysk at $2.14 per barrel, with another $1.50 per barrel for shipping to Mediterranean markets.[97] Thus the netbacks to the producers are expected to be smaller for shipments through BTC, although pipeline tariffs will fall after capital cost recovery.

- In Kazakhstan's onshore Tengiz field, the estimated price needed for profitability is about $15 per barrel. Capital investments in new developments in the northern Caspian region are unlikely at prices of $9–10 per barrel or lower, according to a company source, but producing projects would continue to operate. This cost includes a transport charge of approximately $3.70 per barrel for delivery to the Black Sea port of Novorossiysk.[98] With agreement reached in 2005 to expand the CPC line to 1.3–1.4 mmb/d, future tariff charges are expected to be about $4.00 per barrel,[99] with another $1.50 per barrel for shipping to Mediterranean markets.
- In Kazakhstan's offshore Kashagan field, costs remain speculative, since production has not begun and the development conditions are both difficult and unique. Profitability is likely at a higher level of $15–18 per barrel, in our view, rather than at $12–15. We know that development costs will be steep, and transportation costs an issue, depending on whether a new export pipeline is required. Characteristics making for high production costs are the average depth of 4,300 m (the structure lies "below the salt"), extreme reservoir pressures of 1,000 atmospheres, low API gravity and high content of metals and paraffin, and a high (16–18 percent) ratio of hydrogen sulfide gas contained in the oil. The gas is slated for reinjection utilizing untried and new technology since flaring is prohibited. In addition, the field lies offshore in only 2–10 m of water, and the Sea fluctuates unpredictably, with ice floes of 3–4 m depth blown by the wind in winter. The shallow waters require artificial islands to serve as drilling platforms, and icebreakers and tugs must be designed to limit environmental damage. Islands and icebreakers add to costs. Until expected production levels and their timing become more certain, it is difficult to estimate pipeline tariff costs for existing, expanded, or brand new pipelines.
- In Kazakhstan's onshore Karachaganak field, finding and development costs for full field development are approximately $2.50 per barrel of oil equivalent (this field also includes gas), with operating expenditures (but not including transport charges) placed at $1.80 per barrel of oil equivalent.[100] With a pipeline transport tariff of approximately $2 per barrel to Atyrau and another $2.75 on CPC to Novorossiysk, Black Sea delivery charges amount to just over $9 per barrel. Terminal, port, and Black Sea shipping to the Mediterranean add another $1.50 per barrel.
- In Kazakhstan's onshore fields operated by PetroKazakhstan (before its purchase by CNPC) and LUKoil, production costs were low at roughly $2 or less per barrel, but transportation costs—primarily by rail—east to China or west to join existing pipelines were estimated to fall in the $12–14 per-barrel range.[101] The 2003 opening of a new pipeline serving PetroKazakhstan's operations, howerver, significantly reduced transport charges.

Development costs in this region have a way of rising above their anticipated levels, despite the best-laid plans and efforts to meet cost targets. Costs can be pushed upward by any number of things, including drilling surprises and penalties, fines, or other charges assessed by the host government. These include fines for delay in development (as happened to the consortium developing Kashagan), charges for perceived environmental infractions such as alleged improper treatment of sulfur (as happened to TCO), fines for other environmental infractions including air and water pollution, attempts to require higher pipeline tariffs (as Transneft pressed the CPC consortium), required social expenditures, a variety of other outlays arising from technical challenges (as in the development of Karachaganak, where mercaptans delayed production), and from additional government-mandated expenditures.

As to prices, Caspian producers are price-takers, showing no interest in OPEC membership, which requires adherence to production quotas. Of all the littoral states, only Iran is an OPEC member, and it has yet to begin drilling in its Caspian sector. Caspian countries' output levels are too small to alter OPEC price levels. These producers are free riders under the OPEC price umbrella. Caspian production will respond in periods of high world prices, but will fall in periods when prices do not exceed the estimated $20 real per-barrel costs.

While no one can predict the future level of, or changes in, world oil prices, we can expect price changes and price volatility. In less than a decade, world oil prices have gone from a low of about $10 to a high of more than $70 per barrel. While no one can predict when oil prices will next plunge downward, past history suggests that it will happen. Moreover, we have earlier noted that there is likely to be from 11 mmb/d to more than 16 mmb/d of new oil supplies arriving on the market between now and 2010, including Caspian supplies, which will tend to restrain oil price rises relative to demand growth.

Some observers, particularly the "peakists," who have been maintaining for years that oil supplies should have declined by now, fail to acknowledge that these substantial supplies are already in the "pipeline" and will shortly hit the market. How will these new quantities affect future world oil prices? That will depend not only on the decline of existing producing fields, but also on the growth in world oil demand, which is likely to be driven by economic activity in the United States and China as well as elsewhere in Asia. Should supply rise faster than demand, downward pressure on prices will occur. EIA forecasts world demand in 2010 at 89.7 mmb/d, and world supply at the same level. But CERA forecasts world oil capacity at 101.5 mmb/d, resulting in an excess of nearly 12 mmb/d.[102] If production expands toward capacity, prices are likely to fall. The worry of Caspian producers will be whether such potential price declines will deprive them of profitability from their investments. In a period of low prices, alternative producing areas such as the Middle East (where production costs are roughly $1–2 per barrel, transport is cheaper, and margins are larger) are far more attractive to investors than is the Caspian.

UNFINISHED BUSINESS: DELIMITATION OF THE CASPIAN SEA

The legal status of the Caspian Sea is still unresolved, 15 years after independence. The approach to dividing the Sea remains a contentious issue for some of the bordering states. Since energy resources are not distributed evenly, shorter coastlines for some countries give them smaller shares of the Sea, according to a common method of division. Even if agreement were reached on the principle of division by median line, the issue is where to draw the lines. Each of the littoral states wants its "fair share" of the seabed or, better yet, more than its fair share. Some observers believe that future oil and gas investments will be delayed until there is agreement among all five littoral countries. Others believe that field development and overland pipelines will proceed—and indeed are proceeding—regardless of whether agreement is reached, as long as armed conflict among these countries does not occur. The outbreak of hostilities would almost assuredly result in a slowdown or halt in production, along with investor flight.[103]

Prior to the collapse of the Soviet Union, Iran and the Soviet Union governed the Sea under treaties they signed in 1921 and 1940. Since that time, the dispute has centered on whether the Caspian is a sea or a lake, and what rules might apply to its division and use either by each bordering state or in an agreement for common management.[104] Whether the Caspian is considered a sea or a lake does not really matter. What does matter is how the surface area, with or without common areas, and the seabed are divided. The five bordering states have yet to reach a delimitation agreement that will serve as a foundation for future oil and gas development efforts, and one that all parties are prepared to honor and enforce.

Russia has made the biggest change in its stance on demarcation of the seabed. It initially agreed with Iran that resource development should be deferred until all five bordering states had determined the legal status. It subsequently abandoned that insistence after the discovery of fields that would permit joint Russian–Kazakh development according to a median line division. Once Russia made up its mind to cooperate, it came to agreement quite quickly on Sea division with its neighbors Azerbaijan and Kazakhstan. The agreed division was linked to a given date, and tied largely to river flows and some fairly straightforward mathematics.

Iran and Turkmenistan have followed different paths, although they and the other three seem to have, by mutual agreement, agreed to joint use of the Sea for shipping and fishing, if not as yet for use of the seabed. Iran is concerned that its prospects may be negligible in its 13–14 percent share of the Sea according to median line division. As a result, it is interested in securing a larger share of the seabed. Tehran has long maintained that in any future five-country agreement, each littoral state should receive 20-percent equal shares. The larger Iranian share would reduce Azerbaijan's and Turkmenistan's portions, which is not agreeable to either. Iran now favors a "just" solution and seems to be leaning toward national sectors, but argues that its sector is too small. Tehran made its view known by sending a gunboat in 2001 to drive off a geophysical vessel working at the Alov field in waters it and Azerbaijan both claim. Turkmenistan's issue is where to draw the line, and it has changed its mind several times over the years.

Representatives from the five countries meet regularly but have reached no final agreement. A number of outstanding issues remain between Azerbaijan, Turkmenistan, and Iran. The 18th meeting occurred in October 2005, with no five-country solution on delimitation, despite the usual postmeeting claims of much progress toward resolving disputed matters. At issue are what seabed resources Iran, Azerbaijan, and Turkmenistan claim, and the criteria for how the line is to be drawn. Other hurdles yet to be overcome are issues of ship transit procedures, building of trans–Caspian Sea pipelines, trans-boundary pollution issues, and issues of regional military activity. In terms of Sea delimitation, there are at least four possible outcomes:

- Turkmenistan and Iran could agree to national sectoral division of the seabed along some modified median line principle consistent with the median line agreements of the other three countries, so that the legal regime of the whole Caspian Sea would be agreed and understood. Modified median line division would divide the Sea as follows, with Azerbaijan and Kazakhstan having the largest sectors:
  - Azerbaijan 21%
  - Iran 13.6%
  - Kazakhstan 28.4%
  - Russia 19%
  - Turkmenistan 18%
- The northern sectoral division of the Sea, already accomplished, continues to differ from a southern seabed solution. Iran and Turkmenistan could make their own arrangements, with each continuing its dispute with Azerbaijan.
- No agreement by Turkmenistan or Iran, raising the potential for continued conflicting claims between the two and with Azerbaijan as well, leading to international court proceedings or perhaps to armed conflict.

What are the costs and benefits from agreement or stalling on agreement? The countries could try to secure their preferred portions of the seabed. This course of action will take time and perhaps ultimately fail, or they could agree to a division that would permit more rapid exploitation of any resources found, even if the seabed sector were smaller than each originally had hoped. Since development of some known and prospective fields has been delayed by competing seabed claims, the present value of the income from these fields falls. For Iran in particular the issue is, given the uncertainty of what resource volume it may discover, how much foregone income is it willing to tolerate in exchange for a seabed share of more than the 13.6 percent under median line division? Kazakhstan, Russia, and Azerbaijan all have high discount rates and want to realize the revenues from their shares of the seabed as soon as possible. Iran and Turkmenistan risk being left behind, in the sense that whatever resources already discovered and yet to be discovered in their sectors have a lower present value the longer agreement and subsequent exploitation are delayed.[105] While we believe the latter two are well aware of this tradeoff, the costs have not been high enough to encourage agreement.

It is nonetheless worrisome that failure to agree on where the lines should be drawn in the south could lead to an outbreak of armed conflict there. Although unlikely to turn into a full-fledged or long-lasting war, hostilities may not only physically damage oil and gas facilities, but also cause foreign investors to reevaluate risks and returns. The result would be a slowdown in the size and pace of investment, perhaps a halt for an undetermined interval, capital flight, and further reductions in present value. In the north, Kazakhstan and Russia have no disputes with the southern countries and would likely remain aloof from conflict there. Were conflict in the south to occur after the time when Kazakhstan is likely to be shipping large volumes through BTC, however, Kazakhstan would certainly be an interested party and could be expected to work as an intermediary to settle disputes. While common sense would favor political settlement, there are forces favoring conflict. Turkmenistan's erratic leader might gain some stature at home by focusing on external affairs; Azerbaijan's leadership could demonstrate its resolve and use of its expanded navy to put Turkmenistan in its place; and Iran's conservative government might benefit from a show of force to exercise what it sees as its inalienable sovereignty in the Sea.

PROBLEMS OF THE COMMONS: ENVIRONMENTAL ISSUES

The Caspian Sea and its coastal regions are widely recognized as some of the most polluted areas in the world.[106] Both are well on their way to becoming a completely dead environmental zone.[107] Unaddressed since independence, the Sea's environmental problems have worsened over the last 15 years, despite the 1990s slowdown in industrial activity after the collapse of the Soviet Union (and thus in the volume of waste entering the Sea from the Volga River). Agreement on Sea division could serve as a useful first step to permit needed cooperation and agreement on pollution issues.

Pollution of the Sea and surrounding areas occurs from both anthropogenic and nonanthropogenic sources. Some of these arise from petroleum exploitation, particularly during the Soviet era, while others stem from different types of activity, such as improper waste management. Not all of the pollution is anthropogenic, for the Caspian has the world's highest natural rates of oil and gas seepage. Moreover, the fluctuations in the Sea's water levels also contribute to the spread of that seepage as pollution. In addition, as oil and gas output rises, there is always a chance of pollution from production, transport, spills, and fires. Once Kashagan is producing, a blowout from only one of its prolific wells would constitute an all-Caspian environmental disaster.[108] Some old wells have been leaking for some time, with no steps taken to plug them or to clean up the leaks on the surface.[109] Present and past pollution from oil and gas arises primarily from shore-based wells—which are now nearly depleted—and derelict onshore facilities, with the problems in both exacerbated by bad engineering practices from days gone by and by flooding from the rising levels of the Sea. Tankers have been known to crash into shore facilities as onboard equipment fails, or due to bad weather. Ice and storms are a problem in the winter months in the North Caspian, when moving of drilling equipment becomes more difficult and risks environmental damage. Some pipelines run through earthquake-prone regions, and are at risk of breaks and spill damage along their routes. Finally, there is the question of disposal of impurities, such as sulfur, from the oil production stream. Sulfur disposal is one of several contentious issues between governments and companies, for sulfur must be properly stored to prevent leaching into the water table.[110]

Before the breakup of the Soviet Union, the Caspian Sea's water and air were owned and used jointly by the two surrounding countries, the USSR and Iran. After the breakup, five countries claimed ownership and use of the Sea. Because of the "commons" nature of this ownership, much of the pollution, regardless of the source, is trans-boundary rather than national in scope and impact. Talks on environmental concerns began in 1991, but no country had sufficiently strong incentives to undertake expenditures to improve the environmental purity of the Sea, its shores or air, or to cooperate with others to do so.

The Caspian states now recognize the need for environmental cooperation. At the urging of the United Nations, they signed on November 4, 2003, a "Framework Convention for the Protection of the Marine Environment of the Caspian Sea."[111] The signatories to this document agreed to stop further environmental decay of the Sea and to lower the amount of industrial waste and sewage entering it. The Convention—prospectively the first legally binding treaty on any subject among all five states—will become effective after Azerbaijan's parliament ratifies it in 2006.[112]

The objectives of this highly ambitious treaty are the "protection of the Caspian environment from all sources of pollution including the protection, preservation, restoration and sustainable and rational use of the biological resources of the Caspian Sea."[113] The treaty calls for addressing land-based sources of pollution, seabed activities, vessel pollution, pollution from dumping of wastes, control of invasive sea species, emergency response, restoration and preservation of marine resources, coastal zone management, and research to find ways to deal with the fluctuating levels of the Sea. The principle of "polluter pays" is central to the treaty. This may yet cause problems owing to the trans-boundary nature of the pollutants, where the polluting company or country of origin may be difficult to establish (as is inevitably the case).

While international funding after ratification will help, it will by no means do the whole job, as the task is in the billions of dollars. The parties to the convention have yet to commit to any specific new urban sewage treatments or water purification systems, and it is unclear what enforcement penalties exist if they choose not to pursue environmental improvement or remediation efforts. National action plans have been written identifying the colossal amount of environmental damages needing to be addressed and the concomitant costs.[114]

Unless the states secure the funds and use them cooperatively toward remediation and preservation, the Caspian Sea and its coastal areas will become more polluted. Most of the states understand this "commons" matter, and realize that inattention to it will bring further pollution, degradation, and fishery losses. About 10 million people live in the Sea's coastal areas, and many derive their living from the Sea and its coasts. Unabated pollution will soon threaten not only health, through contamination of the water table, but also livelihoods, through contamination of arable land and loss of fish stocks. Swimming areas are polluted and unsafe, and cancer rates are allegedly on the rise. The consequences may not be limited to environmental or economic ones, but could generate political upheaval as well.

The seabed itself poses environmental challenges, according to some Kazakh scientists. They suggest that earthquakes could be triggered in this seismically active area as the oil is produced from the Kashagan reservoir, potentially affecting both shipping routes and air pollution. They are also concerned that tremors and mud volcanic activity could destabilize the field's reservoirs, compromising its ultimate recovery. Others worry about Kashagan's impact on sturgeon spawning grounds at the mouth of the Ural River, and greater air pollution in the Atyrau region.

Economic Evidence of Pollution: Decline of the Sturgeon and Other Marine Life

Environmental degradation in the Sea has reduced fish stocks, including those of the famed Beluga sturgeon.[115] Pollution has contaminated spawning grounds at river mouths, primarily the Volga, as well as reduced feed for sturgeon. In addition, ships entering the Caspian from the Volga–Don canal introduce new and undesirable fish species through dumping ballast waters; these species limit the food in the Sea for the native species. The Comb jellyfish invasion—arising from the dumping of ballast waters—threatens total eradication of all Caspian fish as it reduces normal fish feed stocks. Unless the five littoral states agree on steps to address pollution or to introduce a second species that will counter the Comb, the livelihoods of all Caspian fishermen are at severe risk.[116]

The Geneva-based CITES in March 2004 threatened a ban on caviar exports from Russia, Azerbaijan, Kazakhstan, and Turkmenistan, noting that those countries had failed to observe the terms of the 2001 Paris Convention on protecting sturgeon.[117] The countries themselves agreed 7 months later to reduce by half their exports of beluga caviar. The United States imposed a ban on caviar imports from all the littoral Caspian states effective September 2005, in the absence of those states providing information as to how they were planning to address sturgeon conservation. In the face of evidence that the 2003 catch levels were less than 1,500 tons, CITES acted on its earlier threat, and in 2006 banned all international sturgeon shipments, at least until Caspian countries provide information on both stock levels and illegal sales.[118] A ban on caviar imports or of trade in sturgeon, however, is unlikely to succeed in dealing with the declining stocks if it is not coupled with means to prevent overfishing and poaching.[119] The lure of high market prices has until now been stronger than the likelihood of getting caught and losing livelihood.

Because the waters of the Sea are currently rising, the risks of pollution from flooded coastal industrial facilities are also rising. The scientific explanation for why the Caspian Sea rises and falls is not complete, but it is largely tied to river flows into and out of the Sea. It is clear that the waters fell until about 1978, when they began to rise again and are still rising. Figure 2.9 shows the precipitous decline of the sturgeon catch over the last 20–25 years, and the current rising pattern of the Sea.

Tradeoffs for the Future: Economic and Environmental Security

These countries will have to face the tradeoff between mounting oil and gas production (and thus growing revenues to fuel potential national prosperity and development) and the risks of new sources of pollution from oil and gas production, transport, spills, and fires. Despite the host countries' and consortia's agreements on the use of world-class environmental standards for drilling, production, and transport, there is always the possibility for incidents of new pollution. With the exception of Kazakhstan's Tengiz and Karachaganak fields, new and future large petroleum developments will

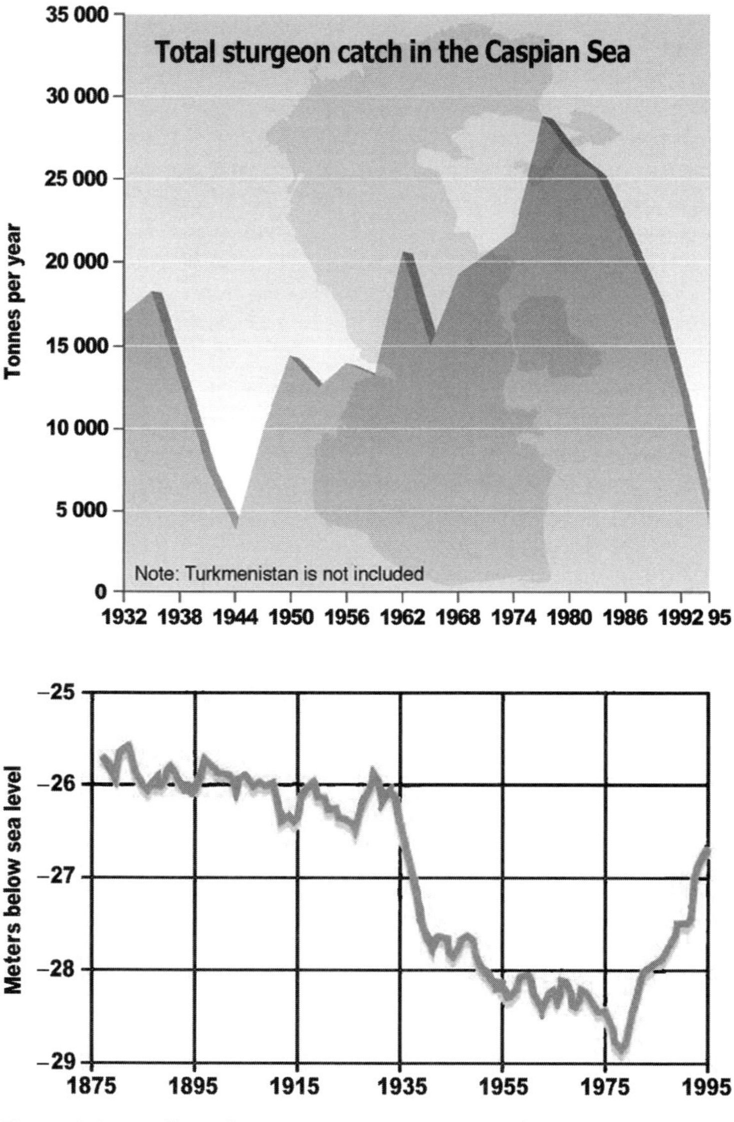

Figure 2.9. Decline of Caspian Sea Sturgeon Catch and Rising Waters of the Sea
Source: UNEP/GRID-Arendal. Reproduced with permission.

occur mainly offshore, where the industry must coexist with fish, bird sanctuaries, and the countries' near-zero tolerance of any pollution from new projects.

Governments have in fact made decisions to move ahead with hydrocarbon development and production. For the most part, they have little else to sell on world markets, and oil and gas is their only major hope for an income stream now and in the future. Fines will be imposed for additional revenues whether the environmental offense is real or imagined. If the host governments push the companies too hard, they may find diminished company interest in the Caspian, given the possibility of potentially more attractive prospects elsewhere.

ENERGY REVENUES: THE KEY TO PROSPERITY OR THE RESOURCE CURSE?

Oil revenues to producing countries in the past 35 years have fluctuated greatly, as both prices and output levels changed. Over time, many countries receiving the revenues have mismanaged them. OPEC in 2005 enjoyed oil revenues of more than $470 billion (in 2005 dollars), although these were well below their peak 1980 revenues (in 2005 dollars) of $572 billion. The question is whether governments this time around will avoid the mistakes of the past, where the receipts were squandered on grandiose spending projects, public debt increased, real economic growth slowed, corruption and rent seeking soared, and vast amounts of money simply went missing. As long ago as 1975, in referencing the problems expected to be generated by oil and gas revenues in 10 or 20 years, Venezuela's Oil Minister and OPEC co-founder, Juan Pablo Perez Alfonso, called oil the "Devil's Excrement."

If resource revenues are well managed by the recipient governments, their size and disposition could well contribute to a glowing future of increased investment and higher economic growth. But this first optimistic scenario is not the only one possible. A second version suggests countries might experience some modest economic growth, but without much dynamism or growth in civil society. A third alternative posits successful exploitation of oil and gas resources, but no economic progress or broadening of civil society. Finally, the fourth and worst possibility is disinvestment and deterioration of the economy, of the social infrastructure, and of civil society, leading to civil unrest and upheaval, as the populace perceives that the favored few have appropriated the rents of oil and gas production.

Oil money permits governments to spend with little questioning or accountability, thus weakening any incentives to reform their economies and stifling any nascent democratic participatory processes. It likewise also removes the necessity to collect taxes elsewhere. The suddenness of the windfalls typically encourages corruption and lack of accountability. The very absence of strong democratic institutions and governments permit these bureaucracies to help themselves to funds as they please for whatever purposes, including remaining in power. These phenomena are generally known as Dutch Disease, a phrase that broadly captures the harmful results of large increases in a nation's income, whatever the source. [120]

Dutch Disease plays out as deindustrialization. The traditional manufacturing and agricultural sectors shrink as the resource sector of the economy booms. If a country catches Dutch Disease, the symptoms can typically be detected in changes in the country's exchange rates and inflation rates. Under flexible exchange rates, as incoming oil revenues are converted into local currency and spent on domestic nontraded goods, the value of the domestic currency rises. This also results in a rise in the real exchange rate. If the exchange rate is fixed, the exporting country's money supply rises, which in turn leads to domestic inflation. In effect this also means a rise in the real exchange rate, as foreign currency now buys less in the domestic economy than previously. In both exchange-rate cases, there has been an appreciation in the real exchange rate that weakens the country's traditional agricultural and manufacturing export sectors. As these sectors shrink, production and unemployment rise. The economy deindustrializes. Labor and capital now look for alternative opportunities in the nontradable sectors such as services.

Evidence on Incipient Dutch Disease: Azerbaijan and Kazakhstan

Both Azerbaijan and Kazakhstan are aware of the risks of Dutch Disease and the actual and potential deterioration of other economic sectors as the resource sector booms. Yet neither is likely to avoid it or contain it. Both countries are overwhelmingly dependent on energy exports, with Azerbaijan's energy sector amounting to more than 80 percent of its exports in 2004, and Kazakhstan's to nearly 60 percent the same year, placing them, according to Professor Paul Stevens, at high risk of Dutch Disease.[121]

Stevens examined countries whose revenues from resources exceeded 30 percent of total merchandise exports as being at risk of Dutch Disease. The higher above 30 percent a country finds itself, the harder it is to deal with the distortions a resource boom induces. While Stevens acknowledges that 30 percent is his choice and is arbitrary, the real issue is what is happening to the rest of the country's traded economic activity, and whether it is dwarfed by the unbalanced growth in the boom sector, not whether the percentage is larger or smaller.[122] After resource depletion, the non-oil activities will form the economic base, and it is those activities that will, or will not, provide future economic sustainability.

Not only are the two countries lopsidedly dependent on export earnings from energy, but they also face the problems of real exchange rate appreciation and growing inflation, both of which are signs of Dutch Disease and work to the detriment of the non-oil sector's expansion. Monetary policy in neither country has been successful in keeping the exchange rate weak and inflation low. Azerbaijan's Economy and Development Minister is already on record stating that in the absence of development of the non-oil sector, Dutch Disease is inescapable.[123] With the prospect for increased oil and gas exports in the future, despite the creation of special Oil Funds in which to place the oil and gas revenues to combat the disease, we believe that Dutch Disease and unbalanced growth are inevitable.

Oil Funds and Their Uses

The Oil Funds in both countries are rising rapidly. At the end of 2005, Azerbaijan reported nearly $1.4 billion, and Kazakhstan just over $8 billion, in their respective Oil Funds.[124] Despite the veneer of oversight, the Funds remain at the personal disposition of the countries' leadership. Historically resource revenues have gone toward pet projects of the leadership or to the privileged few. Use of the Funds for public goods spending, such as for investments in clean water and environmental remediation, health and education, and/or transportation development, so as to build up the non-oil sectors, are not the only alternatives.

The Funds could in whole or in part be distributed to the population at large, where each citizen, with some criteria to prevent fraud and abuse, would receive a certain amount each year as determined by a rule. This method of distribution might go a long way in retaining political support from populations who have experienced little tangible benefit to date from oil and gas activity. Alaska's Permanent Fund works in this fashion, where a resident of the state receives a lump sum payment each year. Distribution to the population as a whole reduces the leeway for the leadership to pledge the flows to an ever-increasing government sector. It also puts money directly in the hands of the populace, and avoids the possibility of government interference or favoritism. It could be managed by the ANB or the NBK. Are either Azerbaijan or Kazakhstan likely to adopt distribution to their populations as a workable alternative use of the funds? We think not, as both governments wish to retain current control of the Funds and make all decisions for future projects.

Other alternatives include EITI and PWYP, both initiatives designed to increase transparency on how much oil and gas money is in government coffers and how it is spent.[125] Both are designed to force accountability and responsible stewardship. The interest in more transparency in government and business transactions is a worldwide phenomenon, not limited to resource-rich developing countries. Both these initiatives, however, illuminate only half the picture, however, since neither focuses on transparency or openness in expenditure programs financed by resource earnings. With no standards for expenditures, corruption and favoritism can flourish, if leaders choose to fund their pet projects.

In summary, Dutch Disease is either already present or a rising risk in both Azerbaijan and Kazakhstan. Until greater transparency of revenue inflows occurs, and rules for selecting spending projects are chosen and adhered to, there is no guarantee that the Funds established to counter Dutch Disease will be successful in meeting the interests of present and future generations. To date, the culture of secrecy, autocracy, and corruption extant on both sides of the Caspian has hindered progress toward transparency. While the countries' hopes for prosperity are completely linked to oil and gas revenues, it is unclear whether these revenues will be utilized for the greater good, or will, as in so many other countries with substantial natural resources, be used to line the pockets of the elite and corrupt.

CONCLUSION

Over the years the euphoria for the size of the Caspian's oil and gas resource potential for production and export has diminished, and the prospects for liberal democracy in this region have likewise deteriorated. We expect Kazakhstan and Azerbaijan to produce together about 2.7 mmb/d of oil in 2010, 3.3 mmb/d in 2015, and just under 3 mmb/d in 2020. Production by others in the Caspian might bring total production up to just over 3 mmb/d in 2010, and approximately 3.75 and 3.6 mmb/d, respectively, in the later 2 years. Full assessment of the resource base, however, remains to be accomplished. These estimates are below those offered elsewhere, and result from our consideration of the frequency of project delays owing to both technical and political developments; these have characterized the exploitation of the resources to date and are likely to continue to do so.

Nonetheless, the interest shown by international oil companies attests to the fact that profits are expected to be made, despite the high costs compared to other areas of the globe, the transportation issues, the legal and environmental problems, and the uncertainties surrounding governments' positions and changes in fiscal terms. The governments of the producing countries, primarily Azerbaijan and Kazakhstan, will confront challenging problems in deciding how to spend the revenues generated by oil and gas. Should their economies become increasingly unbalanced and the leadership spend unwisely, then they will almost assuredly suffer severely from Dutch Disease. The populations will in that case become worse off as living standards fall. In turn, economic deterioration creates fertile grounds for dissidents prevented from expressing themselves to clash with the leadership. Internal unrest and civil war are distinct possibilities throughout the region. Such a climate does little to encourage oil and gas development.

3

The Inner Circle

In this chapter, we review the current and future outlook for Caspian oil and gas production and transport from each of the five littoral states. Given its significance as an energy transporter, we consider Georgia, which while not a Caspian littoral state, is a member of the inner circle. In contrast, we do not include Uzbekistan as a member of the inner circle, for it is of almost no energy or security interest and is also not a littoral state. We include Turkey as part of both the inner circle and the outer circle of countries discussed in Chapter 4, since Turkey will continue to serve as a transport route and as a demand center for both oil and gas.

Over the past 10 years, much of the euphoria regarding the Caspian's inner-circle potential has dissipated. During this period, oil prices fell and subsequently rose sharply to record levels, reserves were reassessed downward, interested parties battled over export pipeline routes, issues of Sea division remained unsettled, a military incident disputing Sea division occurred, state partner firms often faced financial duress and had difficulty providing their capital contributions, and geologic complexities including depth and high pressure proved more challenging than initially anticipated. Even as oil prices rose in more recent years, political as well as economic squabbles continued, with governments trying frequently to get a larger share of the pie and companies claiming that the investment climate is deteriorating and that governments are changing taxes and altering signed contracts on almost a whimsical or capricious basis. Exploitation has also been slowed by lack of offshore drilling equipment and the requirements of drilling in shallow waters, conditions of high wind and ice floes, and the overall need to deal with the environmental fragility of much of the prospective drilling areas.

We examine in detail the oil and gas fields under production or planned for production and the markets for their output. The thrust is primarily directed toward oil production and prospects in Azerbaijan and Kazakhstan, since natural gas markets have for the most part remained underdeveloped, and existing gas pipelines have deteriorated since independence. Azerbaijan, however, will become a gas exporter by 2006 or 2007, once the SCP project to Turkey is completed. Turkmenistan is a major gas producer, and the size of its future revenue stream depends on whether it succeeds in developing gas markets. Leadership issues are also of importance in assessing political stability and in determining whether Western investment will be attracted.

AZERBAIJAN

Azerbaijan borders the western side of the Caspian Sea. Its population is often reported at more than seven million, although as many as two million may have emigrated. According to the *CIA Factbook*, scientists consider the region around Baku to be the "ecologically most devastated area in the world because of severe air, soil, and water pollution; soil pollution results from oil spills, from the use of DDT as a pesticide, and from toxic defoliants used in the production of cotton."[1]

Azerbaijan is one of the oldest hydrocarbon provinces in the world, and has produced oil for more than 100 years. Oil and gas were known to be present in Azerbaijan since before the time of Marco Polo.[2] The very name of the country means "land of fire" in a local language, and derives from Azerbaijan's naturally seeping and burning gas.

Energy History

Commercial refining began in Baku in the first half of the 1800s, and by 1873 about 50 refineries were operating there. Azerbaijan's investment and producing history is generally divided into four periods in modern times: the first oil boom of 1871 to 1921; the second and onshore oil boom from 1921 to 1948; the third and offshore oil boom from 1949 to 1991, and finally the fourth, post-Soviet oil boom from 1991 to the present.[3] By the beginning of the 1900s, with the activities of the Nobel brothers and the Rothchilds, the country accounted for more than half the world's oil production. Output fell during World War I to 47,000 b/d, the same level of output as in 1872. By 1941, however, production had risen to 450,000 b/d, or more than 70 percent of that year's total Soviet production. As World War II continued, output fell to just over 200,000 b/d; the Soviets stopped drilling operations there and moved equipment to Russia's Volga–Urals fields.

Ilham Aliev [AP Photo/AzerTac, Oktay Mamedov]

Azerbaijan: Dynasty Building, Despot Building?

Azerbaijan has not experienced a genuine democratic change of government since its independence. In 2003, the ailing then-president Heydar Aliev named his then 42-year-old son as his presidential successor.[4] Ilham Aliev won the Presidency with 80 percent of the vote. OSCE observers attested to killings, beatings, and many other voting violations. It was apparent well before the election that the younger Aliev was going to win comfortably even in a fair election; it was therefore unnecessary to rig it. The inflows of oil and gas revenues will in our view greatly contribute to his ability to strengthen his political position as President and also provide for a second term in office, beginning in 2008. Two years later, it was an open secret that the 2005 Parliamentary elections were rigged in favor of the Aliev administration. Aliev has indicated no willingness to pursue democratic or political reform and appears to be drifting toward even more authoritarianism than his father's. Corruption is rampant, and dependence on energy revenues nearly total.

In the run-up to his election, and in his inauguration speech, Aliev stressed that he would emulate his father in running the country, and would continue his father's role of preserving "stability." Under the elder Aliev, this meant repeated human rights violations in the name of maintaining stability. In view of the election fraud elevating him to office, Freedom House downgraded the country's rating from Partly Free to Not Free in its 2004 report, and continued to rate the country Not Free in 2005.

Courtesy of his father and a compliant parliament, Ilham Aliev became Prime Minister in August of 2003, but lacked any former significant political experience. He served as titular deputy chairman of the ruling New Azerbaijan Party from 1999, and in 1994 (the year after his father became president) became vice president of SOCAR. He headed the Azeri delegation to the Council of Europe, and was head of the country's National Olympic Committee. Educated at Russia's State Institute for International Relations, he later earned a PhD in history there. His business activities and high living in Moscow and Istanbul during the early 1990s gave credence to his playboy reputation. Despite his attempts in 2003 to change his public image, he continues to be described as a playboy and inveterate gambler, who provoked his father's closing of casinos in Baku in 1998 in view of the size of the son's gambling debts.

At the time of his campaign and election, the Western press labeled Aliev a "puppet on a string," and a "cosmopolitan pseudo-intellectual."[5] With his vast family fortune, however, he has the means to buy off challengers to his authority, and to place favorites in positions of power and money. Nonetheless, his power base is nowhere near as strong or solid as was his father's. According to Stratfor.com, "Ilham Aliev lacks his father's charisma, political skills, contacts, experience, stature, intelligence and authority.... Aside from that, he'll make a wonderful president."[6] Worse yet, an unidentified international banker called him "a complete and total waste of skin."[7] The general conclusion in the aftermath of his election was that he was neither worse nor better than any other candidate, corruption will continue unabated, and the risks of internal instability are far greater now than under the iron hand of the elder Aliev. The 15-year impasse over settling the territorial dispute with Armenia on the Nagorno-Karabakh enclave within Azerbaijan continues, and a resumption of hostilities cannot be ruled out. Should these risks grow, and Aliev be removed from, or die in office, there are no procedures for choosing a successor; the likely infighting among contending groups could lead to a reemergence of civil war.

After World War II the Soviet industry moved offshore, and between 1979 and 1989 the Azeri, Chirag, and Guneshli fields were discovered—these are the ACG fields that the AIOC consortium is now developing. The latest oil boom began after independence in 1991, culminating in the 1994 "Contract of the Century" between the Azeri government and AIOC. Ten years later, as part of the celebrations commemorating the original PSA, the Azeri government, in September of 2004 under the new president Aliev, approved the final development phase of the deepwater portions of the ACG fields. Additional commercial oil fields have not been found, and only one, albeit large, commercial gas field—Shah Deniz—has been discovered.

The Current Players

In total, SOCAR and international oil and gas companies have signed some 29 agreements for both onshore and offshore fields. Table 3.1 makes clear that companies from all over the globe are or have been active, and that they were ready to commit nearly $60 billion to development. Yet the results to date have been disappointing, with only the ACG fields and the Shah Deniz natural gas field amounting to commercial finds.[8] Thirteen projects have either already been dissolved and the participants have relinquished their licenses, or are shut down or abandoned after disappointing drilling results. SOCAR has announced that it will study the development of the fields relinquished, and if further work seems economic, SOCAR or independent companies will begin to do so in 2006–2007.[9] SOCAR has not announced how it plans to finance these projects, and the company has often been short of funds. It now appears, however, that the offshore geology north of the ACG fields on the Apsheron trend is not promising for oil discoveries, while that to the south is primarily one of deep and costly gas.[10] Many companies say privately that there is little more in terms of Azeri hydrocarbons to be found either onshore or offshore.

The southern portion of the Caspian appears to be primarily a gas-prone region, as the discovery of the large non-associated gas Shah Deniz field indicates. The field lies to the southwest of the ACG fields and is likely to be able to more than meet local and Turkish natural gas demand. Its output in the future may actually be constrained by the failure of Turkey's gas demand to grow as originally forecast, unless additional West European contracts are signed. While the search for oil continues, the track record of discoveries in Azeri territory since the country's independence does not inspire geologic optimism.

Production: Present and Future

Azerbaijan in 2005 produced about 440,000 b/d of oil and 4.6 bcm/y of gas (gas data here and elsewhere are for 2004, because 2005 figures were

Table 3.1. Azerbaijan's Field Development Projects, Current and Relinquished[11]

Production-Sharing Agreements	Participants	Percent Ownership	Date Signed	Projected Field Investment	Status
Anshad Petrol (onshore)	SOCAR	51%	1993	N/A	Production of 77,000 b/d in 2004
	Attila Dogan	31.5%			
	Land and General Berhard	17.5%			
Azeri, Chirag and Deep Water Guneshli (offshore)	BP	34.1%	1994	$8–13 billion	Production of 144,000 b/d in 2004; development continuing
	Unocal	10.3%			
	SOCAR	10%			
	Inpex	10%			
	Statoil	8.6%			
	ExxonMobil	8%			
	TPAO	6.8%			
	Devon Energy	5.6%			
	Itochu	3.9%			
	Delta Hess	2.7%			
West Absheron (onshore)	BMB	100%	1994	$700 million	SOCAR takes over 1999 after BMB requests suspension of operations
AzGeroil (onshore)	SOCAR	51%	1995	N/A	Production of 1,000 b/d in 1999
	Grunewald	49%			
Karabakh (offshore)	LUKAgip	45%	1995	N/A	License relinquished 1999
	Pennzoil	30%			
	LUKoil	12.5%			
	SOCAR	7.5%			
	AGIP	5%			

Shah Deniz (offshore)	BP	25.5%	1996	$4 billion	Under development; production to begin in Fall 2006
	Statoil	25.5%			
	SOCAR	10%			
	LUKAgip	10%			
	Total	10%			
	OIEC of Iran	10%			
	TPAO	9%			
Shirvanoil (onshore)	SOCAR	49%	1997	N/A	Production of 57,000 b/d in 2004
	Caspian Energy Group	51%			
Dan Ulduzu-Ashrafi (offshore)	Amoco	30%	1997	$1.5 billion	License relinquished, 2000
	Unocal	25.5%			
	Socar	20%			
	Itochu	20%			
	Delta Hess	4.5%			
Lankaran-Talysh (offshore)	Total	35%	1997	$2 billion	License relinquished, 2002
	Wintershall	30%			
	SOCAR	25%			
	NICO	10%			
Yalama/D-222 (offshore)	LUKoil	80%	1997	$4 billion	2-D, 3-D seismic conducted. LUKoil first well not commercial. No discoveries to date
	SOCAR	20%			
Absheron (offshore)	SOCAR	50%	1997	$3.5 billion	Shut down
	ChevronTexaco	30%			
	Total	29%			

(*continued*)

Table 3.1. (*continued*)

Production-Sharing Agreements	Participants	Percent Ownership	Date Signed	Projected Field Investment	Status
Oguz (offshore)	SOCAR	50%	1997	$2 billion	License relinquished, 2003
	ExxonMobil	50%			
Nakhichevan (offshore)	SOCAR	50%	1997	$2 billion	No commercial discoveries; abandoned
	ExxonMobil	50%			
Kur Dashi (offshore)	SOCAR	50%	1998	$2.5 billion	License relinquished, 2001
	AGIP	25%			
	Mitsui	15%			
	Repsol	5%			
	TPAO	5%			
Southwest Gobustan (onshore)	CNPC/CNODC	80%	1998	$900 million	Under rehabilitation; 2-D seismic ongoing
	SOCAR	20%			
Inam (offshore)	SOCAR	50%	1998	$2 billion	Suspended due to technical difficulties, lack of equipment; BP to redrill
	BP	25%			
	Shell	25%			
Araz, Alov and Sharg (offshore)	SOCAR	40%	1998	$10 billion	Drilling suspended due to Iranian ownership claim
	BP	15%			
	Statoil	15%			
	ExxonMobil	15%			
	TPAO	10%			
	Encana	5%			
Muradkhanli (onshore)	SOCAR	50%	1998	N/A	License relinquished, 2001
	Ramco	50%			
Kursangi-Garabagli (onshore)	SOCAR	50%	1998	$2 billion	Production of 6,600 b/d in 2004
	CNODC/CNPC-HK	50%			

Atashgah (offshore)	SOCAR	50%	1998	$2.5 billion	Shut down
	Japex	22.5%			
	Inpex	12.5%			
	Teikoku	7.5%			
	Itochu	7.5%			
Zafar-Mashal (offshore)	SOCAR	50%	1998	$3 billion	Noncommercial prospects; field to shut down
	ExxonMobil	30%			
	ConocoPhillips	20%			
Lerik (offshore)	SOCAR	50%	1999	$3 billion	Frozen because of lack of investors; lies in waters claimed by Iran
	ExxonMobil	30%			
	To be allocated	20%			
Padar (onshore)	Nations Energy	80%	1999	$140 million	Seismic work ongoing
	SOCAR	20%			
Kalamaddin-Mishovdagh (onshore)	Nations Energy	85%	2000	$50 million	Production of 2,700 b/d in 2004
	SOCAR	15%			
Zigh-Hovsan (onshore/offshore)	SOCAR	50%	2001	$250 million	Delayed by environmental issues; LUKoil pulls out, 2005
	LUKoil	0%			
Pirsagat (onshore)	Sheng Li	50%	2003	$140 million	Rehabilitation ongoing
	SOCAR	20%			
	To be allocated	30%			
Garachukhur (onshore)	Noble Sky	75%	2004	$200 million	Approved May 2005, development pending
	SOCAR	25%			
Binagadi (onshore)	Azen Oil	75%	2004	N/A	Approved May 2005, development pending
	SOCAR	25%			
Kyurovdag (onshore)	Shirvan Oil	49%	2004	N/A	Approved May 2005, development pending
	Caspian Energy Group	51%			

unavailable at the time of writing). Its shares of current world oil and gas production amount to, respectively, 0.6 and 0.2 percent. Its old onshore fields have experienced declines in production, and in 2005 produced only about 35,000 b/d. Despite rapid falloffs in production from these old onshore fields, Wood Mackenzie forecast in 2003 that onshore production could be above 80,000 b/d by 2014, but we doubt that it will reach that level.[12] It is the offshore, however, that is clearly the play of interest. Production from AIOC's offshore ACG fields remained constrained until the BTC main export pipeline opened in 2006. Until then, about 150,000 b/d (called Early Oil Production and begun in 1997) moved through the Western line to Supsa for further transport by tanker through the Bosporus, while SOCAR moved quantities both by rail and through the pipeline to Novorossiysk. Gas production from Shah Deniz awaits completion of the SCP to carry gas to Turkey.

With full operation of the BTC pipeline, the ACG fields' output is expected to rise dramatically, peaking at more than 1.2 million b/d in 2010, as shown in Figure 2.3 (*infra*). But this is at most a short-lived peak, and output subsequently declines rapidly to less than 200,000 b/d by 2024. This means that BTC will have capacity available for production from other Azeri fields, if discoveries occur, or from fields in Kazakhstan from about 2010 forward. The Kazakhs have agreed in principle (but have yet to sign a definitive agreement) to ship from 200,000 to 400,000 b/d on BTC, or perhaps even 600,000 b/d (as indicated by Nazarbayev on the occasion of the inauguration of the BTC line), from about 2012. Barge transport to join BTC is likely to come first; there are hurdles to overcome before a seabed pipeline from Kazakhstan to Baku becomes a reality.

Turkey and Natural Gas: Window to the West

The world is awash in natural gas potential, and Azerbaijan's large natural gas find at Shah Deniz is just another in a large number of gas sources hoping to find a market in Turkey or elsewhere in Europe. Azeri gas faces competition in Turkish and other Western markets from Russian, Iranian, and potentially Turkmen gas. Turkish markets are presently oversubscribed and Azeri gas sales may be deferred for some time.

Shah Deniz is located offshore about 100 km southeast of Baku. It is conservatively estimated to contain more than 400 bcm in non-associated natural gas reserves. Until more contracts are in place, only about 200 bcm are currently considered proven.[13] It also contains some oil, placed at 750 million barrels, but it is under development as a gas field.[14] The first phase of the field's development calls for production of 7 bcm/y. While Azerbaijan and Georgia will buy small volumes (perhaps 200 million cubic meters annually),

Turkey will be the major buyer. BOTAS, Turkey's state gas pipeline company, agreed in 2001 to a take-or-pay contract beginning in 2006. By 2009 Turkey is to purchase 6.6 bcm per year for 20 years.[15] Should demand by Turkey or others warrant phase-two development, Shah Deniz production could rise to as much as 20–30 bcm/y.

From today's perspective, the economics of the development of the Shah Deniz gas field look marginal, particularly if the field faces constrained Turkish demand and suffers in competition against Russian gas priced at roughly $3.00 per MMBtu in Eastern and Central European markets. At this Russian price, the price of delivered Shah Deniz gas to Europe will probably not be competitive. Only a large 20–30-bcm/y pipeline, transiting many countries, offers the scale to deliver competitively priced gas to distant end markets. To commit to a 20–30-bcm/y project, plateau production of at least 20 bcm/y should be expected for at least 20 years, and probably 30 years. The field, in addition, is a challenging one to develop, and no one knows how it will behave once it is—or if it is—fully developed.

Despite the fact that the Shah Deniz field presents drilling challenges[16] in terms of reservoir depth (4,500–6,500 m below the seabed), pressures as high as 14,000 psi, various tectonic stresses, mud volcanoes, debris flows beneath the seabed, and questions of wellbore stability, the main risks confronting gas developers are overwhelmingly those of market economics and the ability of Turkey to honor its contracts. The earthquake in 1999 followed by the collapse of the Turkish economy markedly slowed the country's growth in energy demand, especially for new gas-fueled power plants, and underscored the previous overestimates of Turkish gas demand. Turkey now appears to be significantly overcommitted to gas imports in the years 2006–2008, just when Shah Deniz deliveries are to commence. Gas demand projections have been revised downward by 40–50 percent, but Ankara holds take-or-pay contracts with all its foreign suppliers.[17] These could force Turkey to pay roughly $1 billion annually if it abrogates the contractual volumes, as it has done on several occasions with Iran.[18] Turkey has bought only about one-quarter of its contracted Russian gas imports through the Blue Stream pipeline across the Black Sea. Many analysts believe that for some time to come no new gas import options can be supported, and consider it possible that SCP will be delayed beyond the targeted startup date of late 2006, unless additional sales to Western Europe are won. Turkey is without doubt greatly relieved that the Trans-Caspian agreement it signed in 1999 to buy 16 bcm/y of Turkmen gas through Azerbaijan has become a dead issue, for that pipeline's completion would have made Ankara's natural gas overcommitments even more acute.

For Turkey to avoid take-or-pay penalties, it must connect more of its population to the gas distribution system. In addition, it needs to make sure that the extensions of its gas transport system to Greece and Italy move forward. If these extensions occur, Greece and other Europeans will have the opportunity to buy Shah Deniz gas and could consider buying other sources of Turkey's excess gas supplies. Turkey could also develop gas storage facilities, but these are not expected to be of sufficient capacity to ease the country's overcommitments to gas imports. BOTAS has established gas grids in only four of the country's largest cities, covering only 20 percent of the population.[19] Turkey, however, is already in the process of establishing itself as an international gas bridge to Europe, should BOTAS not need the gas itself.[20] Greece and Turkey signed a protocol of intent to link their countries with a 280-km pipeline, as part of BOTAS's Southern Europe Gas Ring Project. Construction on this line began in 2005. Greece may be interested in buying as much as 5 bcm/y, although initial deliveries will start at less than 1 bcm/y. In turn, Greece and Italy agreed to extend the line 800 km to Puglia, Italy, where it will join the Italian gas pipeline system. While the latter extension is not expected to be completed before 2010 or 2011, Greece is expected thereafter to buy as much as 8 bcm/y. Initial capacity is estimated at a modest 10 bcm/y. This will be a costly line, requiring a deep-sea passage to Italy, and may yet lose out to a cheaper and all-land route to Austria; one such proposal is the Nabucco land route to Baumgarten, Austria. Once one of these extensions is operating, Azerbaijan, Turkey, and Greece plan to expand the line so they can ship up to 30 bcm/y, permitting Caspian gas to potentially reach markets possibly in Italy, but perhaps as well in Romania, Bulgaria, the former Yugoslavia, Austria, and Hungary.[21]

Greece, like other Western European gas-consuming countries, wants to diversify its sources of gas imports, since presently two-thirds of its gas imports come from Russia and the rest from Algeria. DEPA, the Greek state gas company, decided to buy Azeri gas directly from Shah Deniz, rather than buying any excess amounts that Turkey contracted for but cannot use. This will give Greece a market and legal advantage, should the Turks decide for any reason to interrupt deliveries to the west. The real issue will be one of competition on future delivered gas prices (including transit charges). It is likely that Russia can deliver more cheaply to the West. Thus it may make economic sense for Azerbaijan, Greece, and Turkey to seek swaps with the Russians, so that Caspian and Iranian gas reaches the markets of the Balkans, Hungary, the Czech Republic, and Slovakia, while Russian gas formerly serving those countries as well as the Greek and Turkish markets is rerouted to serve European customers further to the West.[22] The current delivered price of Russian gas to Western Europe is over $7.00 per MMBtu, and Gazprom can probably deliver in Central and Eastern Europe for well below that price.[23] Turkey, Iran, and Azerbaijan probably cannot beat Gazprom's delivered price to these areas, but issues of supply security may trump prices.

Risks

The production profiles outlined previously assume all plans are realized fully, and without delays. There are a number of problems, however, that could slow down development and curtail production levels. Programmed development can and often does differ from the size and timing of actual development; this approach of "risking" production is widely used in the literature.[24] We divide these problems into four groups. The first includes those of drilling and geology, which can add to the time needed to complete wells, and thus to well costs, thereby delaying or stretching out the proposed production profile. These include

- Shortage of drilling rigs and other equipment
- High formation pressures (8,500 psi or more), which can result in water inflows into wells, as well as stuck equipment and blowouts
- Deep offshore reservoir depths of from 3,000 to 6,000 m
- Reservoir damage from heavy drilling muds
- Sea water invading the well bores because of erosion around both surface and conductor casings, and
- Mobile mud volcanoes and unconsolidated seabed soils, interfering with both pipelines and platforms.

The second group of problems that could alter AIOC's oil production profile is centered on oil prices. If oil prices should fall much below $12–15 per barrel, which appears unlikely at the present time, company profits will be squeezed and production may not expand as planned. Since oil competes in the consuming market on a net-back basis to the producer, the first component to feel price pressure is the cost at the wellhead. Certainly some of the risks from unanticipated volatility in prices can be reduced through hedging, but hedging does not remove the pressure completely.

The third group of problems pertains to risks, disruptions, and possible natural disasters, foreseen or not, affecting the BTC pipeline and oil shipments through Azerbaijan, Georgia, and Turkey to Turkey's Mediterranean port of Ceyhan. One risk pertains to the source of the volumes that will replace ACG shipments as those fields decline; no firm commitments have been made by Kazakhstan to barge or pipe oil to Baku. If Kazakhstan decides to ship through Iran to the Persian Gulf, or eastward to China, the BTC line may soon become an expensive white elephant in the outlying years. Much of the risk is political in nature, and depends on government decisions as well as private ones. In addition, there are environmental groups in Georgia who claim that the BTC line will kill the mineral water industry in the Borjomi Valley (near the pipeline route and southwest of Tbilisi) should there be a pipeline leak or break; their efforts did bring an alteration in the original route, but the issue of future environmental contamination remains. In Turkey the pipeline confronts both political and natural risk, as it passes through restive Kurdish-populated areas that are also repeatedly subject to earthquakes. While the pipelines are buried, pump and compressor stations are above ground and are easy targets and costly to repair or replace. The BTC partners together with the transit countries are expected to take steps to guard both the oil and the gas pipelines so as to minimize the potential for sabotage and disruption, but there is little defense against earthquakes.

The fourth and by no means smallest concern is the possibility of terrorist attack on the BTC line by any group or individual wishing to make a statement against Western interests and wound "the enemy." There is nothing easier to blow up than a pipeline, as has been demonstrated repeatedly in both Iraq and Colombia. Pipelines are so long and immobile that they cannot be fully protected everywhere. The possibility of renewed war between Azerbaijan and Armenia over the territory of Nagorno-Karabakh, in some areas less than 30 km from the BTC pipeline route, or of hostilities breaking out in Georgia, could also threaten BTC. The line skirts Nagorno-Karabakh to the north, but it lies within a few kilometers of what could become a renewed war zone (see Figure 2.5, *infra*). If war resumes, shipments are likely to be interrupted, as Armenia will seek to deny its enemy a source of funds. These are all wild cards, but cannot be ruled out.

Projected Revenues

Earnings from oil, and in the future from gas (depending on price and sales volumes to Turkey and others), are likely to build rapidly in the coming years. Once the company investors recover development costs, the Azeri government will realize most of the rent, defined as the difference between costs (including returns to the companies) and the sales price. At current record levels of oil prices, the government will realize enormous inflows. Table 3.2 shows estimates of the government's take from ACG, Shah Deniz, and its interest in BTC. The high price estimate was based on a market price of $30 per barrel.

With prices about twice that level presently, and assuming they stay at these levels for some time to come, the government's take is likely to roughly double the amounts shown. In addition to these expected flows, the government of Azerbaijan has already received $300 million from bonus fees on the ACG project, and more than $500 million in profit oil from Early Oil Production. An additional bonus payment of $75 million is due when ACG production rises above 300,000 b/d, which will occur after BTC

Table 3.2. Projected Azeri Oil and Gas Revenues from Production and Pipelines

ACG Full Field Development	Azerbaijan state take (includes SOCAR profit oil, Azerbaijan profit oil and Azeri profit tax):	
1993–2024	$21 billion @ $16/bbl	
	$39 billion @ $20/bbl	
	$58 billion @ $25/bbl	
1994–2024	$107 billion @ $30/bbl[a]	
BTC 2003–2024	Azerbaijan Government	$0.6–0.7 billion[b]
	Georgian Government	$0.6–0.9 billion
	Turkish Government	$1.2–1.8 billion[c]
Shah Deniz (Stage I) and SCP, 1996–2031	SOCAR	$0.5 billion
	Azerbaijan Government	$3.0 billion
	Georgian Government	$0.5 billion

[a] Later estimate (*Sustainability Report 2004*) uses $30/bbl.

[b] In addition, SOCAR will receive $1.8 billion (money of the day [2000]) profit from its 25% shareholding in the BTC project.

[c] Further, BOTAS will average approximately $70 million per annum as its operating fee for operating the Turkish section of the BTC pipeline, on top of the $1.4 billion it gets under the Lump Sum Turnkey Agreement for the construction of the Turkish section of the pipeline. TPAO, with 6.75% equity in ACG and 6.53% equity in BTC, will also benefit from the revenues accruing from these investments, currently estimated at $4 billion and $481 million, respectively.

Source: BP, *Caspian Development and Export, Regional* Review, January 2002, Section 3,4, p. 3, online at http://www.caspiandevelopmentandexport.com/Downloads/RR/Eng/Doc1/RR3.pdf, and BP Azerbaijan, *Sustainability Report 2004*, p. 14, giving estimates for Azerbaijan revenues at $30/bbl over the period 1994–2024, online at http://www.bp.com/liveassets/bp_internet/globalbp/STAGING/global_assets/downloads/A/Azerbaijan_sustainability_report_2004.pdf.

becomes fully operational in 2006. The Shah Deniz bonus fee amounted to $37 million at the time of signing the agreement, and acreage fees of $1,200 per square kilometer per year during the exploration period have been paid. An additional payment of $50 million is due when Shah Deniz production either stabilizes at 12,000 barrels of oil equivalent per day or production from two production wells is "sustained at a stable production rate."[25] The consortium member companies, on the other hand, do not begin to experience positive cash flows until 2006–2007, or perhaps later, because of either technical or market delays.

Conclusion

Azerbaijan is overrated as a potential supplier of both oil and gas. Oil production looks to be constrained by the absence of new finds, and gas production by a lack of robust demand and uncertain international markets. Many concessions have already been relinquished and various consortia dissolved. The country's reserves of both oil and gas are clearly not of Middle East caliber or comparability. The enthusiasm for the new frontier of Azeri hydrocarbons and the view that it will be a major contributor to future energy security are much overstated. Moreover, the political instability and weakness of the rule of law and of democratic institutions raise the risks confronting development projects. Renewed war with Armenia is still a real possibility, and the prospect of opposition to strong-man rule is likely to grow as the population sees little benefit from the inflow of petroleum monies.

Despite the fact that Baku stands to be the major beneficiary from hydrocarbon projects, the interest and hope of the United States that oil and gas wealth will lead Azerbaijan to prosper and stabilize are unlikely to be realized. The government is not democratically chosen (the country has never had an OSCE-blessed fair election), and shows every sign of continuing its grip on power through its control of natural resource revenues and its unilateral decisions for their use. These decisions are made easier by the absence of accountability of the revenues to the public, the absolute discretion of the President on the use of the Oil Fund, and the absence of penalties if the Fund is indeed misused. With hydrocarbon revenues providing the bulk of governmental revenues, the temptation to spend may simply become too great to resist.

GEORGIA

Slightly smaller than Azerbaijan, Georgia is a state of about 4.7 million people that borders the Black Sea but not the Caspian Sea. Georgia's importance to Caspian energy development is its role as a transportation link between Azerbaijan and Turkey, and for that reason we consider it a part of the inner circle. It has not discovered any significant indigenous resources of oil and gas, although it has conducted exploration offshore in the Black Sea. Once a relatively rich part of the Soviet Union, it received low-cost Soviet energy. As Georgian political and trade relations with Russia soured following independence, Russia reacted with higher-priced energy sales, tipping Georgia's economy into freefall. Ever since, Georgia has suffered from economic stagnation, decay, and widespread poverty, civil strife, crime and corruption, regional secession, and the absence of the rule of law. While Georgia remains dependent on Russia for most of its oil, gas, and electricity, it sees BTC as a means of reducing that dependence.[26]

Mikheil Saakashvili [AP Photo/Fritz Reiss]

Georgia: Rule by Roses?

Led by Mikheil Saakashvili, Georgia's Rose Revolution in November of 2003 forced the resignation of one of the longest-serving Soviet-era leaders, President Eduard Shevardnadze.[27] Presidential and parliamentary elections followed; Saakashvili, then 36 years old, won 96 percent of the vote and took office in early 2004. He and his supporters won on a reform and anticorruption campaign, but well into his second year as president, his popularity was down precipitously to as low as 30 percent, as citizens saw little improvement in economic conditions. Half live below the poverty line, and unemployment approaches 20 percent.[28] Encouraged by what it perceived as progress toward democracy, the United States in 2005 awarded Georgia a Millennium Challenge Award.[29] The Saakashvili regime has cracked down on corruption and smuggling, with those guilty forced to make payments to the state to secure their release. Illegal detention and torture continue, however, and Saakashvili has yet to implement his promised enhancements of democratic standards and economic reforms. Some see a growing climate of fear, as individuals are increasingly unwilling to speak their views.

Georgians' expectations for the new regime were and are high. The economy is in shambles and the secessionist provinces remain defiant. If the new leadership can reestablish Georgia as a state, there may be a chance for democracy to take permanent root. In foreign affairs, Saakashvili's twin policy goals are to establish good relations with Russia while maintaining and deepening ties with the West. Accomplishing both may be difficult.

Saakashvili is the first leader in the region to have received his education in both the former Soviet Union and in the United States, and to have lived in both the United States and in Europe. He is a graduate of the Kiev International Relations Institute in Ukraine, and studied at the Strasbourg Human Rights Institute in France. He won a scholarship to Columbia University and obtained a master's degree in law. He later enrolled in a doctoral program and studied at the George Washington University. He subsequently worked at a New York law firm and at the Norwegian Human Rights Institute in Oslo before returning to Georgia in 1995 to enter politics.

Elected to Parliament, Saakashvili served as chairman of the parliamentary committee for constitutional and legal matters. Thereafter he became Georgia's representative in the Parliamentary Assembly of the Council of Europe. He is one of the first leaders in the region whose political career has no roots in the Soviet period. Shevardnadze appointed him Minister of Justice in 2000, but he resigned a year later, accusing Shevardnadze and members of his government of massive corruption. Thereafter he founded an opposition party, the National Movement, and entered Tbilisi municipal politics, becoming a member of the Tbilisi local council and in 2002 its chairman.

Young and inexperienced, Saakashvili is viewed by some as brash, volatile, unpredictable, headstrong, unwilling to take criticism, and likely to make mistakes. In particular, he lacks the sophisticated political maturity and judgment demonstrated by his late partner in government, former Prime Minister Zurab Zhvania.[30] He tends to be confrontational in his approach, and whether he has the competence to govern is an open question. Under his leadership, Georgia is now a one-party state, with the vast majority of parliamentarians supporting Saakashvili, and no opposing views aired.[31] He is, nonetheless, politically acute and succeeded in harnessing public frustration with the abysmal economy and high levels of corruption to catapult himself into office. Some critics suggest that his consolidation of power may be paving the way for establishing a future dictatorship. If lawlessness continues and/or an attempt on his life is successful, Georgia is likely to slide into complete chaos that may threaten the oil and gas pipelines as well as provoke an as yet unknown Russian action.

Energy Pipelines for Economic Prosperity

The country is pinning its hopes for economic development in the longer term on the energy corridor from Azerbaijan to Turkey, and possibly on the future development of its own oil and gas fields.[32] The country's main source of future income is likely to derive from the transit fees it will earn from the BTC and SCP oil and gas pipelines, respectively.

Absent the risks of instability and disruption, Georgia is expected to earn from the BTC oil pipeline 12 U.S. cents per barrel in each of the first 5 years, rising to 14 cents per barrel over the next 10 years, and thereafter to a minimum of 17 cents per barrel.[33] During the period 2006–2010, if production from the ACG fields builds as anticipated, oil transit revenues should build from $6.8 million in 2006 to $46 million in 2010 and $51.1 million in 2012.[34] Thereafter, these revenues are expected to fall to about $10 million annually by 2020, unless additional Kazakh volumes are shipped. Georgia already receives oil transit fees of $0.19 per barrel for the Western pipeline carrying about 150,000 b/d of Azeri oil from Baku to the port of Supsa; on average this amounts to $8.7 million annually.

With regard to natural gas, Georgia's take from the SCP pipeline from Shah Deniz to Erzurum in Turkey is expected to amount to $0.67 per mmcf ($2.50 per thousand cubic meters) in each of the first 5 years, and rising to $0.134 per mmcf ($5.00 per thousand cubic meters) in the later years. Initially Georgia will take its tariff payments in kind, and would thereafter buy up to 500 million cubic meters per year at an average price of $1.48 per mmcf.[35] On the basis of SCP throughput beginning at 2 bcm in 2007 and rising to 8 bcm in 2012, Georgia would earn from $10 to 44 million each year, according to the International Monetary Fund.[36] Total transit fees for both oil and gas (including fees on the Baku–Supsa line if that line remains in service carrying current volumes) thus amount to from $15.5 million in the early years to as much as $104 million in 2012. Transit revenues thereafter will depend on the volumes of oil and gas shipped to Supsa and to Ceyhan and Erzurum. These are not insignificant sums for the budget of a country that is at present struggling to develop sources of revenue other than donor aid, value-added taxes, and customs duties.[37] The former Georgian energy minister noted that over the 40-year lifetime of BTC alone, Georgia is likely to receive $2.5 billion in oil transit fees, and potentially a larger amount in gas transit fees.[38]

Both pipelines' futures link the common interests of Azerbaijan and Georgia, since both are depending on pipeline revenues as sources of income. Conspiracy theories abound here as in most former Soviet states, with news reports and some officials noting the possibility of either Russian sabotage of the lines, or of attacks by criminal or terrorist elements, or by radical environmental groups, with or without Russian support.[39] Clearly unrest in either country makes the challenges of building and operating either line even greater, and raises the risk as to whether the revenues will be realized in the amounts forecast.

Guarding and providing pipeline security on the 150-mile segment in the country will be the task for the Georgian military, with assistance from the U.S. military through the train-and-equip program. The Georgian government also signed a contract with Northrop Grumman to develop an aerial radar surveillance system to monitor the pipelines and the adjacent area on the Georgian segment.[40] This monitoring would be used to alert Georgia to both hostile incursions as well as to environmental events. But it is the responsibility of the BTC pipeline consortium to pay for both pipeline security and environmental protection. The company has agreed to provide $165 million for these purposes over the 20-year life of the project.[41] Nonetheless, interference with pipeline transport is easy.

Conclusion

Georgia's future existence as a state is uncertain, but the Saakashvili government will continue its attempts to reunite the breakaway provinces into its national "constitutional space" after its peaceful ouster of the leader of Ajaria and its overtures to both Abkhazia and South Ossetia. The BTC and SPC pipelines offer both financial promise and energy supplies. The leadership cannot afford undermining of the pipeline projects through political unrest or sabotage, nor can it do without the future revenues these lines will generate. The issue remains as to whether and how the government will ensure smooth functioning of both pipelines. Moreover, if the government succeeds in its aggressive campaign against corruption, the transit revenues may have a chance of being well spent.

KAZAKHSTAN

After Russia and Iran, Kazakhstan is by far the most well-endowed oil and gas province in the Caspian region. In addition to its resources, the country's population (more than 15 million) and geographic size dwarf both those of Azerbaijan and Georgia. Kazakhstan has shown strong economic growth, primarily due to its petroleum sector. In 2005, Kazakhstan produced more than 1.2 mmb/d of oil from all fields, and 18.5 bcm of gas (gas data for the year 2004).[42]

Nursultan Nazarbayev [AP Photo/Ivan Sekretarev]

Kazakhstan: Postponing Democracy

The Rose Revolution in Georgia and the Orange Revolution in Ukraine caused consternation in the Central Asian republics, where leaders do not countenance removal from power by popular opposition and a free voting process. In parliamentary elections held in September 2004 (moved up from the original date of December 2004), Nursultan Nazarbayev's Otan party won handily. OSCE election observers found that the elections were flawed and did not meet international standards. His elections have been shams, with no meaningful opposition. Nazarbayev won an easy and rigged reelection to the presidency in late 2005. During his years as president, he has become an increasingly autocratic and authoritarian leader. Political pluralism is not to be tolerated.[43]

Nazarbayev earned his political spurs in the Soviet regime, and has been the only president of the country. Born in 1940, he joined the Communist Party at age 22. His education includes a technical degree in metallurgy and a PhD from the Russian Academy of Management. Before moving into Party political circles, he worked at a steel mill from 1960 to 1969.[44] Thereafter he did Komsomol and Party work, rising to become president of the Kazakh Soviet Republic. At independence, he ran unopposed for president, receiving 98.7 percent of the vote.[45] He won gratitude from the West with his early steps to remove and return all nuclear weapons to Russia. Nazarbayev is an astute politician who believes order and reform must originate from the top. A corruption proceeding against Americans charged with bribery of high Kazakhstan oil officials, allegedly including the President, is ongoing in New York. He has repeatedly stated that his country is not prepared for democratic practices, and that what it needs is stability and continued economic growth.

In 1995 he extended his rule to the year 2000, but then called an early election in 1999, which he won handily. His decision in mid-2005 to spend millions on pensioners, students, and public sector salaries demonstrates his maneuvering to buy loyalty ahead of his next presidential election,[46] which he moved up a year to December 2005. The parliament remains quiescent, and abolished the two-term limit for the presidency as well as the age limit. Those government officials offering criticism are normally quickly fired, and press and other critics are either beaten, arrested on trumped-up charges, or receive strong messages to cease their criticisms.[47] Pinning its hopes for stability and wealth on future sizeable oil and gas revenues, the government is at the same time souring the climate for foreign investors by introducing new regulations on foreign-funded energy projects, by redefining the interpretations of signed agreements, and by imposing penalties for a variety of ever-changing offenses, including environmental violations.

Corruption is as pervasive in Kazakhstan as in other countries in the region. The elites' drive to enrich themselves is a contagion. In the absence of a functioning and accountable democracy, which is looking ever less likely, this contagion/tradition will continue. Should opponents of Nazarbayev gain office, or were his daughter to succeed him, no dissimilar behavior is expected, for this will be their chance to help themselves.

But oil and gas resources alone do not automatically turn a land-locked country into a net exporter contributing to energy diversity and world energy security. There must as well be favorable political conditions, pipelines to connect resources with markets, and functioning oil and gas markets. World oil markets exist, but gas markets for Kazakh supplies are as yet limited. Moreover, most of Kazakhstan's gas is associated gas, meaning that it is found mixed with oil. Thus the country's gas production is linked to its oil production, and the costs are likely to be high for installing gas treating and gathering facilities. If Kazakhstan's only export outlet for gas is sale to Russia at below-cost levels, most of its gas is unlikely to be developed. Proposed new long-distance gas pipelines to China are costly, and some of these markets will face competition from LNG deliveries to China's industrial east that may be cheaper than long-distance gas pipeline shipments terminating in China's west.

With hydrocarbon reserves estimated by DOE/EIA at 9–17.6 billion barrels, and even higher by BP, Kazakhstan is the dominant resource holder in the Caspian region, and is expected to dwarf production from all its Caspian neighbors. Three large fields—Tengiz, Karachaganak, and Kashagan—contain 70 percent of recoverable oil reserves and nearly 75 percent of recoverable gas reserves.[48] Smaller and mostly onshore fields account for the remainder of reserves. While there are unexplored and as yet untapped offshore prospects, these have yet to be tendered for drilling.

Many projections have been made for future levels of Kazakh oil and gas production. With the exception of Tengiz, most of the interest focuses on the country's offshore prospects. Natural gas markets have yet to be developed (other than for sale of Karachaganak gas to the gas-processing plant across the border in Russia's Orenburg). For now, most Kazakh gas is slated for reinjection to maintain pressure in oil wells. Estimates of future production include those made by the government and by an outside consultancy:

- In May of 2003, Kazakhstan's President Nazarbayev issued a decree approving the development plan through 2015.[49] This program called for total oil production to reach 1.2 mmb/d by 2005, 2.3 mmb/d by 2010, and 3.5 mmb/d by 2015. Most of this growth is to come from the three fields named above. According to the development program, investments of $11 billion will be required by 2005, another $20 billion in the 2006–2010 period, and finally $21 billion in the period 2011–2015.
- A presidential decree, however, does not ensure commercial development, even though the 2005 goal was reached in 2004. Major oil companies doubt that these plans to increase production in the out-years will succeed, given the limits on ownership by foreign investors in any project, the new law that imposes greater taxes on PSAs, and the success of the government in taking a share of Kashagan.[50]

- The consultancy Wood Mackenzie estimated in 2003 that Kazakhstan had the potential to produce as much as 1.9 mmb/d by 2007 and about 2.3 mmb/d by 2011, depending on whether the ambitious development programs were met. But Wood Mackenzie flagged the difficulties of arranging financing and transport, as well as other technical difficulties, commercial delays, and the concerns for fiscal and political stability. As a result, the company cautioned that a realistic actual production level may differ from a potential production estimate. While production potential could be as much as 1.9–2.0 mmb/d by 2007, actual production may be well below the government's estimate of 2.3 mmb/d in 2010.[51] By way of comparison, we project 1.850 mmb/d by 2010.

Kazakhstan is a net natural gas importer, primarily from Uzbekistan, which supplies about 4 bcm/y in the south of Kazakhstan. Much of the gas produced in the west of the country in association with the country's oil production is slated for or actually used for reinjection, since flaring is prohibited. In Kazakhstan's south, the Amangeldy gas field is under development by the state oil company and Repsol-YPF. The hope is that this field's output will replace imports of Uzbek gas in the future. Wood Mackenzie estimated that Kazakhstan's natural gas production potential could be around 40 bcm/y by 2012, but realistically reduced that estimate to around 30 bcm/y because of development delays, lack of growth in demand, and the absence of hard-currency export markets.[52]

The Resource in Detail: Tengiz, Karachaganak, and Kashagan

Private company interest in Kazakhstan's oil and gas prospects has been strong in the years since independence. The *Oil & Gas Journal*'s proved reserves for Kazakhstan held steady from 2003 to 2005 at 9 billion barrels of oil and about 1,840 bcm of natural gas. Whether these estimates are accurate is less important than whether the reserves are exploitable in both an economic and political sense, and whether competitive market outlets exist. Figure 2.6 (*infra*) shows the three major fields under development: Tengiz, Karachaganak, and Tengiz.

Tengiz

This field and its adjacent structure, Korolev, north of Tengiz, were discovered in 1979. Both lie onshore to the east of the northeast shore of the Caspian Sea, and both are currently producing. ChevronTexaco, the operator of the joint venture TCO, estimates Tengiz recoverable crude oil reserves at 6–9 billion barrels, and associated or solution gas at 400 bcm. Korolev may have 2–3 billion barrels of oil, with producing facilities connected to those at Tengiz; Korolev gas figures have not been made public.

Formed in 1993, TCO includes ChevronTexaco with 50 percent, Exxon-Mobil with 25 percent, the Kazakh state oil and gas company KazMunaiGaz with 20 percent, and LUKArco with 5 percent. The contract provides for a 40-year primary-term[53] license area at Tengiz (from 1995 until 2035) covering 1,600 square miles, and a concession at Korolev until 2030. Production has increased from 30,000 b/d in 1993 to about 320,000 b/d in 2005.[54] Total investment in the field and associated facilities to date is more than $9 billion, with an additional $21 billion committed over the life of the field.[55] The project is believed to be the single largest foreign investment in the former Soviet Union.

TCO recognizes that Tengiz is a geologic and technical challenge to develop.[56] It plans to more than double output to 700,000 b/d by 2010, despite the difficulties. The field is the deepest under development in the world, with its producing oil zone beginning at a depth of nearly 4 km. The oil is under high pressure, has as much as 28 percent hydrogen sulfide content, and also contains paraffin and corrosive mercaptans (sulfur-containing organic chemicals).[57] The associated gas is also laden with both hydrogen sulfide and carbon dioxide. To meet market demand for low-sulfur oil, the consortium processes it to remove sulfur and other impurities. Acres of sulfur blocks sit adjacent to the field, but there is no market for this amount of sulfur. In the absence of gas markets, TCO plans to reinject sour gas to maintain pressure and increase oil output to its 700,000 b/d target by 2010.[58] The reinjection plan depends on new technology, as yet unproven. These are indeed ambitious expansion plans, and their realization will be closely tied to the consortium's relations with the Kazakh government and the availability of pipeline capacity and transportation routes.

The absence of high-volume transport facilities limited output from Tengiz and Korolev until 2001, when CPC opened. For years TCO shipped oil by rail and barge, as well as by pipeline through the Russian system. TCO was the world's largest rail shipper of crude oil until CPC began operations. Current plans call for all Tengiz production to move via CPC, which will expand to 1.3 mmb/d by 2010. Not yet clear is whether all of this oil, along with production by other oil shippers on CPC, will transit the Bosporus to international markets, move through a Bosporus bypass, or move northward on the Odessa–Brody line, should the current direction of flow on that line be reversed.

Karachaganak

Karachaganak is another major field that ChevronTexaco and others are developing. Like Tengiz, it is an onshore field to the north of Tengiz and close to the Russian border. Discovered in 1979, it remained underdeveloped until after the dissolution of the Soviet Union, although pilot production of gas condensate began in 1984 and was shipped to Orenburg in Russia for processing. In 1992, BG and Agip received the exclusive rights to negotiate for the field's redevelopment. In 1995 Kazakhstan, Gazprom, and BG and Agip/ENI signed a PSA, but without KazMunaiGaz ownership. Thereafter ChevronTexaco bought a 20 percent share from BG and Agip/ENI (the operators, who now each own a 32.5 percent stake), and in 1997 LUKoil purchased Gazprom's 15 percent interest.[59] The consortium, known as KIO, has a 40-year license from 1997, which KazMunaiGaz monitors.

The field is estimated to contain about 2.3–2.5 billion barrels of API-light recoverable oil and condensate and about 450 bcm of gas. While a new 650-km pipeline links the field to the CPC pipeline at Atyrau, initial shipments fell nearly a year behind schedule in 2004 because of technical difficulties (paraffin content and contamination by mercaptans). Additional challenges to production include the reinjection of high-pressure sour gas to maintain reservoir pressure, and the wide swings in surface temperature. Once the technical problems were overcome, first shipments began in the second half of 2004 at more than 200,000 b/d. Total oil condensate production is targeted to reach 240,000 b/d by 2010.[60]

The Karachaganak development program has experienced the strain of Kazakhstan's independence from the Soviet Union, for Russia now views gas from this field as a competitor to its own. Gazprom's original 15 percent holding in the field was the price for processing the gas at Orenburg in Russia and then exporting it. Since LUKoil took over the Gazprom interest, KIO lost interest in selling gas to Russia. Current plans call for building a $600-million gas processing plant on site at Karachaganak, which will free KIO from continued dependence on Russia as a buyer. But until the new gas-processing plant is built and new gas pipelines are constructed to supply internal or international markets, it is likely that KIO will be forced to sell from 5 to 10 bcm/y of gas to Russia at relatively low prices. Figure 2.7 (*infra*) shows the relevant pipeline and field locations.

Kazakhstan is pursuing a strategy of trying to increase domestic use of natural gas and at the same time avoiding dealing with Gazprom, the Russian monopolist. Since most Kazakh gas production is associated gas, however, it is difficult to separate it from the oil stream and use it. Once separated, gas can either be reinjected to maintain pressure for oil production, or it can be processed and cleaned prior to shipment to end users. Project economics in each case are likely to determine the outcome, but it appears that Kazakhstan will not accept the risk of reduced oil production should gas not be reinjected but sold elsewhere. The irony is that Kazakhstan is both an exporter and importer of gas to and from Russia, generally paying far more for its processed imports than it receives from the sale of its raw, sulfur-laden exports. If the country could move its gas from western fields to demand centers in the east, it could reduce its gas imports from both Russia in the north and from Uzbekistan in the south, thereby presumably reducing prices to consumers in both regions. Growth in gas demand, however, is still sluggish, and financing for pipelines connecting the gas-producing regions in the west to the eastern consuming regions has yet to be found.[61]

Kashagan

The Kashagan field is the newest of Kazakhstan's large oil discoveries, located offshore south of Atyrau and west of Tengiz. Discovered in 2000, the field was declared commercial in 2002. Reserves are estimated at 7–9 billion barrels, although gas injection may raise potential proved reserves to 9–13 billion barrels. BP apparently considers the field's reserves to be much greater, since its *2005 Statistical Review* carried the proved reserves for the country (not the field) at 39.6 billion barrels. Since the field is new, however, no one really knows how much it contains in terms of recoverable reserves; it is still under assessment. In many respects, the field's characteristics, including the API lightness of its oil, resemble those of Tengiz. The target peak production level is 1.2 mmb/d for the years 2015–2016 through 2025, as shown in Figure 2.4 (*infra*) provided there are sufficient export options.

In 1993 Kazakhstan invited exploration bids for its sector of the Caspian Sea. Agip and other consortium members conducted seismic and environmental studies. In 1997 the consortium negotiated and signed a PSA covering some 6,000 square kilometers. Agip KCO is now developing the field. Agip, the operator, and its four major partners—Shell, Total, BG, and ExxonMobil—held equal shares of 16.67 percent each, with ConocoPhillips and Inpex holding 8.33 percent each, until BG decided to sell its share in 2003.[62] The owners' decision to exercise their preemptive rights to buy, when any of the partners chose to sell, set off a storm of controversy with the Kazakh government. The latter announced in mid-2004 that it planned to take an ownership share in Kashagan equal to the BG interest, asserting that all subsurface preemptive rights lie with the government.[63] Parliament passed ex post facto legislation in December 2004 giving such rights to the state, allowing it to use them both prospectively and retroactively. At the end of January 2005, the consortium and the government reached a compromise whereby the government, through KazMunaiGaz, bought half the BG interest, valued at $1.23 billion.[64] The other consortium partners bought the remaining BG interest.

The 2004 legislation raises questions about the sanctity of signed contracts and capricious changes in the rules, and about whether the government may insist on priority to buy into any project now or in the future. These uncertainties may yet set back the timetable for Kashagan development that has already slipped 3 years from 2005 to 2008. The law has likewise increased the squabbling among the partners, who would have preferred to have had one less partner so as to facilitate decision making. Agip KCO has other issues with the government as well, for it is facing a probe as to whether it is properly paying value-added tax and customs fees.

First production of Kashagan oil of 75,000 b/d is now expected in 2008, although the date could easily slip to 2010. The ambitious development plan for Phase I calls for production of 450,000 b/d by 2010. Phase II development calls for 900,000 b/d by 2013, building to 1.2 mmb/d beginning in 2016.[65] The Phase I target will depend on overcoming high reservoir pressures, 3-mile reservoir depth, 15 percent sulfur content, carbon dioxide, associated gas that is prohibited from being flared, extreme weather and ice conditions, extremely shallow (less than 4 m) but fluctuating sea levels, and marine and other environmental sensitivities. In view of the government's plans to move oil east to China, to send it west through the CPC and BTC pipelines, and to double the size of swaps to the south with Iran at Neka, some observers doubt that even with Kashagan's sizeable future output that there will be sufficient oil produced to realize all these ambitious targets. This will certainly be the case if Kashagan is substantially delayed. Our judgment is that production from Kashagan will be delayed beyond 2008, perhaps for up to 2 years or more.

A key issue for Kashagan production levels is transportation alternatives. Only large anticipated production potentials justify the high capital investment for new pipelines. The art is to know how to time production to use existing lines instead of constructing new ones, and to know when and where to build new ones as future increasing production merits. While the U.S. government may favor multiple pipelines, commercial considerations alone generally do not. Yet no company or country wants to be at risk of having only one export route that might be subject to political or technical interruptions. All of Kazakhstan's existing pipeline export routes cross Russia. In 2002 the two countries signed a long-term accord guaranteeing Kazakh oil transit of at least 350,000 b/d, although this is insufficient to accommodate future output from Kashagan.[66] Moreover, Kashagan partners are wary of dealing with the Russian state oil pipeline monopoly, Transneft, for fear of Transneft-imposed increases in transit fees or discounts in crude prices, or competition from Russian companies seeking pipeline space.

As a result, Agip KCO owners are looking at a variety of transportation alternatives for Kashagan production. All have disadvantages that may limit the field's production potential if sufficient transport space is not available when needed. Pipeline transportation charges alone are estimated at $5–7 per barrel,[67] with additional barge or rail charges pushing costs even higher. These costs affect profitability, which is clearly dependent on the expectation of continued relatively high oil prices. Proposals for moving Kashagan oil include:

- CPC west through Russia to Novorossiysk. Currently at capacity throughput of 565,000 b/d, CPC is slated to expand to 760,000 b/d by 2006, to more than 1 mmb/d by 2008, and to 1.3 mmb/d by 2010. Further expansions have not yet been planned. An expandable pipeline link between the offshore location of Kashagan and the CPC line at Atyrau would also be required. Bosporus oil tanker issues again become important, as the Turks fear the environmental risks of ever-increasing tanker passage. A key problem is that none of the Kashagan owners have rights to future CPC capacity; to ship on CPC they would have to buy capacity rights from CPC.

- Shipment through Russia to its port of Primorsk on the Gulf of Finland. The Baltic Pipeline System and the Primorsk port's export capacity were expanded to 1.2–1.3 mmb/d in 2004.[68] Primorsk is subject to winter ice conditions, but these are not as bad as those at Novorossiysk. Another disadvantage of Primorsk is that tankers are limited to 150,000 tonnes given the concern for safe passage around the Danish peninsula. Since Baltic ports are a long way from Caspian fields, tariffs are likely to be high at $7–9 per barrel.
- Shipments through Ukraine and Russia on the Odessa–Brody pipeline, with further delivery to ocean ports at Gdansk and/or Wilhelmshaven, once connecting pipelines are built. No limitations on tanker size have been imposed at those ports, and they are ice-free in winter. We judge this route as likely in the future, as Ukraine wishes to discontinue shipments of Russian oil south from Brody to Odessa.
- Barge or pipeline transport across the Caspian to Baku for shipment on BTC. Kazakhstan ships about 40,000 b/d from Aktau to Baku for further rail shipment to Batumi in Georgia.[69] By their ownership shares in BTC, Agip KCO participants have rights to 150,000 b/d of BTC's export capacity. This could rise if AIOC yields some of its capacity should it not be able to fill the line. While the press has from time to time mentioned shipments of 200,000–600,000 b/d of Kazakh crude on BTC, no specific commitments have been made.[70] Timing will be critical, for Azeri ACG production is expected to fill the 1.2-mmb/d pipeline until about 2010. If ACG's production profile proceeds as planned, BTC could carry additional shipments of less than 100,000 b/d in the years 2012 through 2016, rising to 200,000 b/d and more in the years following. But in view of the large production profile anticipated for Kashagan, BTC capacity alone may not be adequate.
- Barge or pipeline transport of about 40,000 b/d across the Caspian to Makhachkala for further shipment through Russia to Novorossiysk. In addition to the problems of maintenance and Chechnya issues, this pipeline cannot handle large volumes and does not offer a quality bank at its terminus. Russia had hoped to expand this line to 1 mmb/d as a means of derailing the construction of the BTC route; that strategy failed, and the line was not expanded.
- Northward through the Atyrau–Samara line, joining the preexisting Russian system. The Samara line is currently relatively small, with a capacity of only about 320,000 b/d. Plans to expand it beginning in 2006

to 500,000 b/d[71] would provide some potential space for moving future Kashagan production.

- Barge transport of up to 160,000 b/d to Iran ports and swaps for Iranian oil at Persian Gulf export hubs. While this method of export is already occurring, despite U.S. policy toward Iran, the issue is whether the Iranian system can and will be expanded sufficiently to permit movement of as much as 1 mmb/d.
- For Phase II of Kashagan, which is designed to lift production above 450,000 b/d in 2011, some authors note the idea of looping the existing CPC line and connecting it via a spur to BTC through Abkhazia in Georgia. The cost of a spur is estimated at about $5 billion. This line would add flexibility as it avoids the bad-weather delays in tanker loadings at Novorossiysk as well as congestion and delays at the Bosporus. Georgia, however, has yet to open discussions with BTC about such a link, and the turbulence in Abkhazia means this proposed spur has a very low probability.
- A de novo large pipeline, either through Iran, eastward to China, or southward through Turkmenistan and Iran. Any of these proposals are a long way from realization, given U.S. policy toward Iran, the unpredictability of the current Turkmen regime, and the costs of construction to China. Kazakhstan has kept alive the idea of a new export pipeline to China. In 2005 the $700-million, 1,000-km Atasu leg was completed. The line may ultimately deliver 200,000 b/d to the Chinese border. Pipeline length and capacity are to be expanded in later years. At this point the Kazakhs are counting on Russian oil to fill half the capacity of the initial line. The Russians, however, have not yet indicated their willingness to do so, and are leery of China's potential ability to dictate oil prices once the line is completed.[72] While Nazarbayev reiterated in mid-2004 his preference for a crude export pipeline through Iran to Persian Gulf export terminals, it appears he now favors the line to China.
- Given the experience of TCO, the Kashagan partners know rail transport remains an option. This method is the least preferred of all the choices, however, since it is the highest cost. Nonetheless, observers expect from 30,000 to 40,000 b/d to be shipped by barge to Baku for further rail transport to Georgian ports, and an additional 60,000–80,000 b/d to move by rail to various other destinations.[73]

Our judgment is that no new dedicated export pipeline for Kashagan production will be built for the first phase of development, and it is doubtful whether a new pipeline is needed for the second phase, given the plethora of alternatives. Nazarbayev recognizes that a new pipeline through Iran will not happen until political relations improve between the United States and Iran, and the volumes are sufficient to justify the investment.[74] We expect Kashagan production to be delayed long enough to find some capacity on the BTC line, perhaps as much as 400,000–600,000 b/d, with the remainder moving by a combination of the other existing/expanded routes mentioned above. Multiple routes diversify the producers' risks, even as they may raise costs.

The author's conversations with company representatives from the consortia operating at Tengiz, Karachaganak, and Kashagan all lead to one conclusion: these are costly and high-risk projects, in both the technical and political senses. Given the historical roller-coaster of oil prices, profitability is extremely sensitive to world price levels and to government tax takes, often based on revenues rather than on profits. Moreover, there is always the risk that what Western companies thought were inviolable contracts can be changed arbitrarily at the will of governments.

The Investment Climate

The investment climate in Kazakhstan has cooled noticeably over the last several years as distrust has grown between the foreign investors in the consortium and their government counterpart. The reason for this is that Kazakhstan believes it offered too generous terms in the first round of developments, and now wants adjustments in those terms. It appears to be pursuing a policy of deliberately stiffening the terms for foreign investors, tightening the tax clauses, and making new contracts less appealing.[75] The government's announcement—and subsequent parliamentary approval—that it intended to buy a share of Kashagan is an example of one of the frictions between the government and foreign investors. The new tax terms, effective on contracts signed after January 1, 2004, significantly tilted the scales in favor of the government, raising its take from 65 to 85 percent of oil income. As expected, private investors considered the terms too severe, and have expressed dissatisfaction and discouragement with them. The changes effectively reduce the private rate of return to 10 percent on new offshore projects and will apply to all foreign investors. According to ExxonMobil, "the rate of return should be about double (that) to make offshore projects more attractive in the Caspian Sea's difficult weather conditions."[76] Subsequently Shell said it needs at least 15 percent to cover the risks in exploration and development, as well as in transportation and pricing.[77] Companies fear continued reinterpretation of contract and tax terms, and also of lawsuits that appear to contravene the terms of their contracts and serve as a transparent attempt to produce more government revenue by any means. Since the 2004 tax revisions, few new agreements have been signed, and the companies' ardor for Kashagan has diminished.

TCO and Agip KCO have had major differences with the government, as has PetroKazakhstan, a Canadian company (recently sold to CNPC) whose smaller projects are located in the center of the country. TCO has repeatedly clashed with the local and national Kazakh governments. In late 2002 TCO suspended for 3 months the $3.6–4.5-billion development of its sour gas injection project after the Kazakh government objected not only to the consortium's accelerated amortization plan (which would have reduced government revenues), but also to the interest rate associated with the $2 billion in loans arranged by TCO to cover a Kazakh budget gap.[78] The TCO suspension set back the plan for doubling production by about 8 months to a year.[79] The dispute was settled when TCO agreed to help KazMunaiGaz arrange financing for its $700-million share of the costs. Another issue was the government's levy of a $70-million fine on TCO for improper sulfur storage, which TCO maintained was done according to world-class environmental standards. The fine was eventually reduced to $7 million, once the expansion financing was arranged for KazMunaiGaz. In September 2004, the Kazakh government also pressured TCO into agreeing to pay $810 million more in taxes for the 2004–2006 period than it had expected according to the terms of the original PSA.[80] All of these disputes have increased the level of rancor and mistrust between the government and TCO.

The development plan for Kashagan, like that for Tengiz, has not been free from controversy. The consortium submitted its development plan in January 2003, but the Kazakh government failed to approve it until more than a year later in February 2004. The delay curtailed the willingness of the consortium to commit capital, since the project is likely to require as much as $30 billion in investment over its lifetime. Despite its own delay in approving development, the government imposed a fine of $150 million on the companies for delaying first output by 3 years to 2008 ($50 million per year for each year of delay), since any postponement affects the timing of revenues and taxes to the government.[81] This level of annual fine is expected to continue should production be delayed beyond 2008, as we expect. Making matters even worse is the existing confusion as to whether the PSA permits penalties to be levied for delay; typically PSAs do not. Industry analysts believe that the Kazakh authorities are using the issue of timing as a means to renegotiate contract terms so as to win a higher tax take. KazMunaiGaz's ownership in Kashagan as a partner will provide revenues to the government earlier than previously, for the PSA states that only when the partners have earned a reasonable rate of return does the government's share of the profits rise.

Drilling conditions not only at Kashagan, but also elsewhere in the North Caspian, are extremely difficult. Wells are estimated to cost more than $100 million each, among the most costly in the world.[82] The oil lies some 4,500–6,000 m below a thick layer of salt, and requires the use of expensive drilling equipment. High pressures and high temperatures, together with sour gas, add to drilling costs. Sulfur must be removed and hydrogen sulfide gas disposed of. Other environmental concerns loom large, as artificial drilling islands are required to protect the ecosystem, shallow waters have insufficient depth for dispersion of spills and pose risks of supply boats running aground, and high winds and ice floes slow drilling activity in the winter. The risks are great, and must be compensated by an appropriate rate of return.

The problems of PetroKazakhstan, which produces about 150,000 b/d, were relatively new. The company experienced repeated difficulties with both its partners and the Kazakh government. These may have led to the willingness of the shareholders to sell the company to CNPC for $4.2 billion, a premium of nearly 25 percent[83] over share price closings. The sale to CNPC was finalized in October 2005, with CNPC agreeing to sell one-third of its purchase to Kazakhstan, which had threatened to block the deal under the 2002 revised regulations (which permit the state to buy a 50 percent share). PetroKazakhstan was an integrated producer at the Kumkol fields in central Kazakhstan, and the output of its Shymkent refinery met about half the country's demand for oil products. It and LUKoil Overseas were equal partners in Turgai Petroleum, and at the time of sale to CNPC were embroiled in a dispute on pricing and transportation, crude oil production and marketing, and performance of obligations, actively suing and countersuing each other. The isolation of PetroKazakhstan's fields from markets had imposed high transport costs on the company. Despite lifting costs among the lowest in the world at less than $2.50 per barrel,[84] the company's rail costs per barrel were estimated at $14–14.50 prior to the opening of the KAM pipeline in 2003.[85] Now that CNPC is the primary owner, it is likely to favor moving its production from central Kazakhstan to the east, given the recent completion of the Atasu pipeline to China.

Prior to its sale to CNPC, PetroKazakhstan experienced growing troubles with the Kazakh government, problems that were potentially far more damaging than its disagreements with LUKoil or its dependence on costly export routes. The Kazakh authorities forced the company to cut daily production by 30 percent to comply with the prohibition of gas flaring. The company was also in its second lawsuit against the government, which accused it of violating antimonopoly legislation. In the first case, it paid a fine of about $3.6 million, and it faced a second fine of more than $55 million, which it planned to appeal. The government filed criminal charges against two of the company's expatriate executives. The issues in both cases were collusion, and that the company raised products prices without the approval of the state, earning unlawful profits of some $6.8 million that must be returned to the state. The company claimed that its contract permitted it to price its products on a commercial basis and that the antimonopoly agency had no power to interfere, to set prices, or to require the company to sell oil at reduced prices to farmers during the planting season. The company was in Kazakh courts on this matter and on a criminal case along the same lines at the time of the CNPC bid.[86]

Foreign investors were disturbed by both the government's attack on PetroKazakhstan as well as the uncertainty of changes in Kazakh laws affecting future production. Although existing contracts have not been scrapped, new ones have far more onerous terms, and are subject to changes in the tax code, in the petroleum transfer pricing and investment laws, and in the requirements governing tenders for services (local content), and work permit regulations. Both the new tax code and the transfer pricing legislation hinder foreign investor activities, and the investment law challenges the preexisting provisions. Specifically, the tax regime would no longer be unchangeable for 10 years in foreign contracts, the protections against asset expropriation are reduced, and the provisions regarding international arbitration in case of disputes are weakened. The new provisions on tendering/procurement rules and work permits are motivated by the need to develop Kazakh service and manufacturing companies and provide jobs to Kazakh workers, but such rules tend to raise costs.[87]

The January 2004 tax provisions were adopted after less than 30 days' consideration, and with no input from industry. The American Embassy in Almaty reported that the new legislation significantly raised the burden on the oil and gas sector, and it agreed with analysts who suggested that investment will be discouraged and the development of the industry slowed.[88] A new export tax will apply to all future non-PSAs, and has the goal of creating budget stability for the government. The particular provisions for cost deductibility are not clear. Since the tax is based on revenues rather than profits, the Embassy suggested its implementation could result in tax liabilities exceeding actual revenue. Projects are already slowing down or being postponed as a result of the new tax clauses that have affected projected profitability.[89] There is also a fear that the success Moscow had in bringing down Yukos may encourage Kazakhstan to hit companies with tax arrears to make up for what it now perceives as its overgenerosity in the earlier rounds of granting licenses and PSAs. Uncertainties in contract terms, surprises in laws and regulations, and uncertainties regarding the court system have dampened foreign investor interest in Kazakhstan.

The companies' resistance to the tax increases is turning into recoil, which may lead to their future unwillingness to engage in new projects, and perhaps even walk away from existing ones. The Kazakh government appears to be giving little credence to Western notions of contract sanctity. Its decisions have also delayed startup of exploration projects with Russian companies for the same reasons, who also question the fiscal soundness of the proposed agreements. The Kazakh government believes it has a natural right to a share in Kashagan since the resource lies below its seabed share. Having already invested more than $2 billion through 2003, and prepared to invest $10 billion by 2008, the private partners may be loath to walk away from Kashagan, but this action cannot be ruled out. Such a development would leave Kazakhstan with very little development offshore and would hurt its nascent contractor and service industries, as well as diminish and delay the flow of revenue to the government. The government does not seem to recognize or care that it risks losing foreign interest, and thus future production, should it continue tightening the terms or fail to provide stable arbitration and court systems to keep foreign investors actively engaged.

Other Kazakh Prospects and Proposed Developments

Despite the coldness of the investment climate, the country believes it has the resources and that they will be developed in a timely fashion to meet its targets. As noted previously, Kazakhstan issued in 2003 a presidential decree, the goal of which was to increase oil production to 2 mmb/d by 2010, and to 3.5 mmb/d by 2015, with that level to be maintained over the next quarter century. More than a presidential decree will be necessary to reach this level, and it will not be realized without huge sums invested and transportation facilities completed. Required investment is put at more than $40 billion through 2015. The decree and the Program for Development further note that natural gas production is likely to amount to 60 percent of total output and to contain 19 percent sulfur, thus requiring either sulfur removal or reinjection for pressure maintenance. The Program flags the need for the country to develop its internal gas utilization strategy as well as plans to export it.[90]

Few international observers would agree with the Program's calculation that offshore recoverable reserves in the Kazakh sector of the Sea alone (not the country as a whole) amount to 58.4 billion barrels.[91] Even if one adds total Kazakh oil reserves of 9 billion barrels carried by the *Oil & Gas Journal* or 39.6 billion barrels carried by the BP *2005 Statistical Review* to similarly sourced gas reserves of 65,000 bcf or 11.7 billion barrels of oil equivalent, that total is below what the Program reports for the offshore sector only. These numbers are simply neither believable nor supportable. Skeptics conclude that if reserves are overestimated by this great a factor, future production and its timing are likewise overestimated. Shell, who is a participant in Agip KCO, publicly expressed doubt that Kazakh production will reach 3 mmb/d before 2015.[92]

Foreign companies have been waiting for some time for new opportunities in the Kazakh offshore sector, but many uncertainties remain. Very few additional details on the actual implementation of the Program for Development are available, and it is not clear which blocks will be offered for participation with KazMunaiGaz. Companies are reexamining how the new tax rules will affect their returns. Moreover, the disposition of the vast amounts of natural gas together with the size of investment for production and transport of either gas or oil are likely to slow down future developments.[93] According to the American Embassy and the U.S. Commercial Service, the

Kazakh government has identified 120 potentially promising offshore structures, and it plans to offer three or more for tendering each year beginning in 2006 through 2010, thereafter adjusting the rate as needed.[94] KazMunaiGaz also has the right to grant contracts instead of arranging open tendering. If it proceeds with block-by-block contract negotiations, companies may reassess the uncertainties that come with lack of transparency. At long last, however, after consideration for more than 2 years, Nazarbayev in mid-2005 signed a new law to govern offshore PSAs, which spelled out that KazMunaiGaz is to have not less than 50 percent of any new offshore project.[95] While there will no doubt be disputes on bonuses and other terms, the new law's terms should offset in part other uncertainties that have plagued the offshore investment environment prior to the law's passage.

While no one knows which blocks will be offered and what the terms and performance requirements may be, reporting from the U.S. Commercial Service indicates that the Program considers that in addition to Kashagan, such structures as Kurmangazy, Kalamkas-more, Aktoty, and Kairan are ready for drilling.[96] Agip KCO is also the holder of the offshore block in which the Kalamkas-more prospect is located, while the Aktoty and Kairan prospects lie in its Kashagan block.

For projects straddling the median line of the Sea, the Russians have expressed interest in joint development with Kazakhstan. The two countries agreed in 2005 on the joint development of Kurmangazy and Khvalinskoye, two of three such prospects. The third is named Tsentralnaya. All three have suffered delays due to reduced profitability under the changes in the Kazakh tax code. Kurmangazy is in the northern part of the Sea, and is to be jointly developed by Rosneft, Zarubezneft, and KazMunaiGaz; the PSA was signed in July 2005. Recoverable reserves are estimated at 135 million barrels of oil,[97] although estimates of probable recoverable reserves of oil and gas range up to more than 7 billion barrels.[98] If the field is declared commercial, first production is not expected for at least 10 years. Khavalinskoye, primarily a gas field further south, is a 50-50 joint venture of KazMunaiGaz and LUKoil, signed in 2003. That field is estimated to have oil reserves of 73 million barrels and gas reserves of 127 bcm, much smaller reserves than those of Tengiz, Karachaganak, or Kashagan. If the field is declared commercial, production is to start in 2010 or later. In addition, Gazprom, LUKoil and KazMunaiGaz have teamed up to consider development of the Tsentralnaya prospect offshore Makhachkala, with the two Russian companies creating a joint venture. Recoverable oil and gas reserves in this block were placed at 3.8 billion barrels, with gas amounting to 75 percent of that total.[99] All three prospects, however, have been plagued by scarcity of financial resources, the absence until mid-2005 of a Kazakh PSA law governing offshore development, by requirements to comply with Kazakh local content and educational regulations, by disagreements on the location of service industry bases, and by changes in the tax environment.

LUKoil's interests, however, are not limited to prospects that lie on either side of the median line. In early 2004, it finalized an agreement with KazMunaiGaz for seismic surveys and exploration drilling in the Tyub–Karagan and Atash structures of the northern Caspian Sea. These areas lie fully within the Kazakh sector northeast of the Khavalinskoye prospect. LUKoil's first test well at Tyub–Karagan, in water depths of 5–15 m to a depth of 2,500 m, found no oil. Unknown at present are the size of reserves (the estimates vary), the sulfur or other impurities, or the costs of lifting and delivering to international pipelines.[100]

While there are many smaller fields and prospects throughout Kazakhstan, the interests of both the companies and KazMunaiGaz are likely to converge on those with larger estimated reserves, and on those whose development costs and transportation costs can be minimized. Future pipeline or barge tariffs may be key, for that part of the cost may often determine whether the oil is left in the ground. Future offshore fields in the Kazakh sector of the Sea will be assessed in terms of whether output can be linked to pipelines that carry Tengiz and Kashagan production, so as to minimize transport cost.

Conclusion

Kazakhstan has promise as an oil and gas province, but in our view it is oversold in terms of the size of its resources and it is underrisked through understatements of the magnitude of industry-related and political challenges. The identification of producible oil reserves occurs only with investment, and investors evaluate where they place their funds on the basis of considerations of risk and competing alternatives for the allocation of their capital. The key to whether any of these current and future reserves are exploited and commercialized depends on the expected market price of oil, the cost structure, and perhaps most important of all, the political environment, which in turn influences the investment climate. Three major fields hold most of the oil discovered to date, and it is simply speculation if not fantasy to add unproven reserves figures from undrilled structures.[101] Natural gas reserves are less of an issue as to accuracy in size, since for some time they are likely to be used for reinjection to maintain pressure in wells producing oil rather than to be marketed.

Regardless of the geologic or economic size of Kazakh oil and gas reserves, it will likely take years to set up adequate regulatory, legislative, and governance mechanisms, even under the best of circumstances. But Kazakhstan and KazMunaiGaz have yet to fully adopt the rule of law, which means that caprice, bribery, tax uncertainty, and uncertainty about regulatory and judicial interpretations of contractual terms and operational norms are major concerns to foreign investors, and raise their risks and costs of doing business there. Weak and/or corrupt institutions tend to discourage investors; the latter need stability and predictability in a known and transparent investment climate, together with known and reliable mechanisms for resolving disputes. Yet in the last few years investors have expressed concerns that previously approved contracts would be forcibly redrafted by a government that believed it was overly generous in the past. If the executive branch can interfere whenever it chooses, it imitates Soviet times and shows that investors' capital is not protected by contracts under the law.[102] The country continues to be governed from the top down by the president, himself a product of the former Soviet system, who tolerates no opposition or argument from the parliament. The energy industry remains a cash cow for the government, whose main interest is keeping itself in power.

What does this assessment mean for world energy security and from the perspective of the United States? There is no possible way Kazakhstan is ever going to displace either Russia or Saudi Arabia as the world's leading oil exporter. Costs are high, and production, engineering, and political difficulties loom large. We expect Kazakhstan to produce less than 1.7 mmb/d by 2010, and perhaps 2.6 mmb/d by 2015, well below the presidential decree goals of 2 and 3.5 mmb/d, respectively, for those dates. At the margin, another million or so barrels per day of output matters, in that it works to retard any rise in price, but we view official Kazakh optimism as unwarranted. Moreover, in the years ahead increases in Caspian production are likely to be more than offset by declines in North Sea output and elsewhere. Not all of Kazakh production will be exported, of course, since rising domestic demand will absorb some of it. Forecasts for internal use, however, depend on economic growth, and both the demand forecasts and growth forecasts can and do vary greatly. While individual consortium companies may yet prosper from their Kazakh investments, the fruits of their labor will in no way change the fact that the comparative resource advantage lies in the Middle East.

TURKMENISTAN

Turkmenistan, a country of about 5 million people, has substantial reserves of natural gas. Since the fall of the Soviet Union it has had no strategic importance whatsoever to the United States because of the psychological instability of its leader, President Saparmurad Niyazov.

Turkmenistan is looking to market its gas wherever it can, either north through Kazakhstan and Russia, south through Afghanistan to Pakistan, or east to China. While its oil production is still small, increasing volumes will require an oil export pipeline. The leadership seems to live in a fantasy land in terms of the size of its reserves of oil and gas, and its projected production levels for both.[103]

Saparamad Niyazov [AP Photo/Alexander Zemlianichenko]

Turkmenistan: One-Man Rule for Life, and Then?

Also a product of the leadership and power structure in the Soviet Union, Saparmurad Niyazov has been dictator of Turkmenistan since 1985. He ran unopposed in the first popular presidential Turkmen election in 1990, and was overwhelmingly elected. In June of 1992, he became president of independent Turkmenistan. Two years later, he extended his term to 2002. In 1999 the People's Council amended the constitution to name him president for life.[104] He says he will step down in 2010 at age 70,[105] and that he will name a candidate to stand for the presidency in elections in 2009. As part of his effort to promote his cult of personality, he declared himself "Turkmenbashi," or father of all Turkmen, and insists on being addressed as Turkmenbashi the Great. *The Washington Post* lists Niyazov as one of the world's 10 worst dictators.[106] Whatever change occurs when he dies or steps down is likely to be unpredictable and chaotic; there is no chance of a "flower" revolution. Niyazov's actions and decisions seem to indicate his intent to take his country back to the Stone Age; Turkmenistan under his repressive rule is a failed state.

Born in 1940, Niyazov grew up in an orphanage. He earned a degree in power engineering at the Leningrad Polytechnic Institute in 1966, and worked thereafter at a power station near Ashgabat. A member of the Communist Party from 1962, he rose to appointment in 1985 as the country's Chairman of the Council of Ministers, was subsequently elected to the top party post in Turkmenistan.

Niyazov's control over the economy is total, and there is no transparency of budget decisions. While Turkmenistan regularly holds sham elections, opposition candidates are excluded. The makeup of the parliament does not really matter, since it has no real power. The President makes all the decisions and can and does ignore parliament. He is a total tyrant known for his "dictatorial eccentricities."[107] Increasingly the words *megalomania* and *lunacy* are used in descriptions of his rule.

Turkmenistan could become a significant and prosperous energy producer and exporter, particularly of natural gas, should it establish terms attractive to foreign energy investors. So far it has failed to do so. The U.S. Embassy in a sharply worded cable in mid-2003 flagged the hostile investment climate, noting that there is no established rule of law to protect property and contract rights, that there is excessive, inconsistent, and non-transparent regulation, the currency exchange system is severely restricted, and the government is largely unfamiliar with international business norms and practices.[108] The United States and others have also repeatedly criticized the country's human rights abuses. Because of his totalitarian control of the political and economic environment, foreign investors have been either noticeably reluctant to invest in energy projects, or have completely lost interest in doing so, especially since Niyazov has stated that the oil and gas industry will not be privatized until at least 2015.[109]

Niyazov is believed to be failing in health, and perhaps seriously ill.[110] He does not travel outside the country for medical treatment any longer for fear of a coup. There is no apparent successor to Niyazov. Were he to die or be ousted in a coup, we expect a turbulent interval. The growing repression and isolation in Turkmenistan, combined with its declining living and rapidly deteriorating health standards, suggest the possibilities for serious infighting and even civil war when Niyazov leaves the stage.

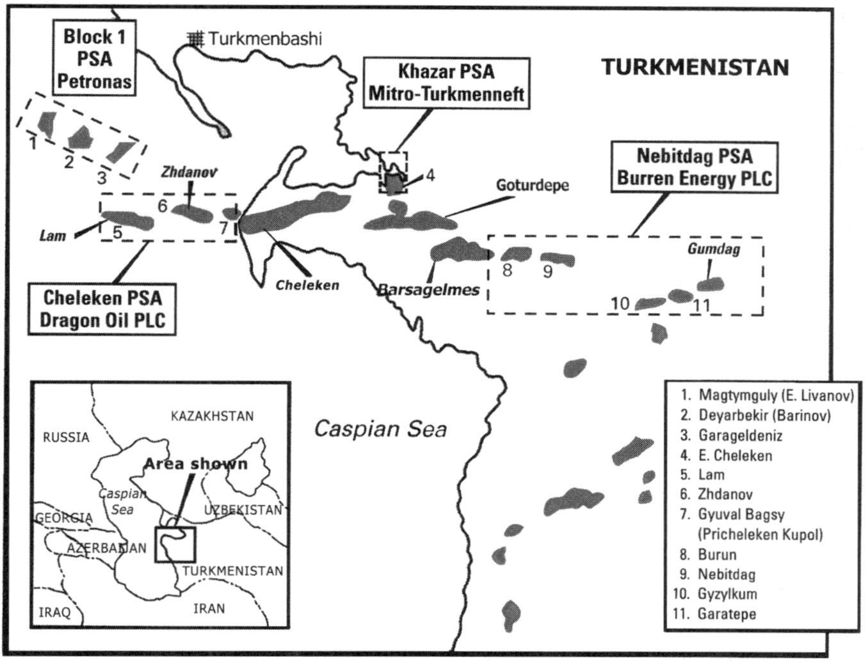

Figure 3.1. Turkmenistan Oil and Gas Fields in or Near the Caspian Sea
Source: Mark McCafferty and Valentina Kretzschmar, "CPC, BTC pipelines make current Caspian area oil export capacity adequate," *Oil & Gas Journal*, Vol. 102, Iss. 18, May 10, 2004, 52 (7). Reproduced with permission.

Locked in the Past

President Saparmurad Niyazov has led the country since independence, and controls the workings of the state in every way, despite the existence of a Parliament. The arbitrariness and capriciousness—some would say lunacy—of his rule have kept the populace ignorant, repressed, and poor, and have discouraged foreign investors in the energy sector. Nonetheless, the country is well endowed with natural gas, amounting to some 2.9 trillion cubic meters. Eighty percent of its commercial energy reserves are gas.[111]

Turkmenistan has two hydrocarbon basins, the Caspian shelf and the center/eastern region, where the latter extends into Uzbekistan.[112] The inland region is primarily gas-prone, and contains little oil. Figure 3.1 focuses on the Caspian Sea area in the west where oil fields as well as natural gas deposits are located. The state oil company, Turkmenneft, has signed only five PSAs with foreign companies, despite its announcement of ambitious plans for energy development.[113] In this hostile investment climate, few foreigners are willing to commit capital.

Oil and Gas Production Targets

While the country would like to exploit both its oil and its gas, Ashgabat has failed to agree on division of the Caspian Sea. It also overplayed its hand regarding the proposed Trans-Caspian oil and gas pipeline project, which is for all practical purposes dead. That project was intended to carry from 25 to 30 bcm/y from Turkmenbashi to Baku and onward to Turkey and to other West European markets. The project failed for a variety of reasons, including Niyazov's demands for an up-front payment rumored to be as much as $400 million, the inability of Azerbaijan and Turkmenistan to satisfactorily divide the property and shipping rights through the line, and uncertainties about the future size of Turkish and other Western gas markets. With the successful completion of the Blue Stream natural gas pipeline from Russia to Turkey and the AIOC decision to build the SCP, the Trans-Caspian project was upstaged, receded into the background, and abandoned.[114] The other failed alternative was the proposed 1,680-km Trans-Afghan Pipeline from Turkmenistan's large Dauletabad field near the Iranian border southward through Afghanistan to Pakistan and India.[115] Unocal, now part of ChevronTexaco, was one of the original project sponsors, but abandoned the idea in 1998 in the face of the Taliban regime's activities in Afghanistan. As a result, Turkmenistan was forced to continue selling gas to Russia and Ukraine at monopsonistic prices, since there are no alternative pipeline export routes for Turkmen gas, other than a small line to Iran. These Turkmen failures provided Azerbaijan the opportunity to move ahead with a shorter and more economical gas pipeline from Shah Deniz to the west, while at the same time avoiding serious earthquake risks likely to affect the Trans-Caspian pipelines proposed in the central Caspian Sea.

Turkmen production potential has been greatly overestimated. As far back as 1994, Turkmenistan's annual production goals for 2000 were 560,000 b/d for oil and 130 bcm/y for gas.[116] The country's actual production in 2000, however, was nowhere near that level, amounting to 144,000 b/d of oil and 43.8 bcm of gas. In 1994 the Turkmen government also forecast that in the period 2010–2020, oil production would rise to 1.6 mmb/d and gas to 54.6 bcm/y. Oil production in 2005 was just under 200,000 b/d and gas production amounted to 55 bcm/y (2004 data), and neither is likely to meet the 2010–2020 predictions,[117] although the prospects for increased production appear to be better in the gas arena. The problems include the following:

- The targets of the 1990s were unrealistic. The Turkmen government had little comprehension of costs, project financing, and timing of production.
- The country is landlocked, and there was then and is now no way to move such volumes, even if producible, in the absence of new transport facilities and of greater willingness on the part of Russia to ship Turkmen gas.
- There is no oil export pipeline; oil that is exported is barged either to Neka in Iran or to Baku in Azerbaijan.

In November 2003, Turkmenistan again announced that between 2003 and 2020 the country would increase its oil production to 2 mmb/d and its gas output to 240 bcm/y.[118] According to the *Petroleum Economist*, "If achievement matched ambition, Turkmenistan would be one of the world's largest oil and gas producers."[119] It is neither. Observers give these targets no more credence than they did to the ones offered 10 years ago. In fact, there is some doubt whether Turkmenistan is going to be able to meet all its contractual agreements for gas sales to Russia and Ukraine, which together amount to nearly 140 bcm/y by 2020.[120] Until additional oil fields are found and developed with international technology and capital, total oil output is unlikely to rise to even 1 mmb/d by 2020.

Turkmenistan also announced in late 2003 that it will "require" investment of $63 billion by 2020, presumably to meet its announced production target levels. Of this amount, it expects to secure more than $25 billion from foreign investment in offshore PSAs in the Sea, with the remainder to be provided by the state-owned oil and gas firms through revenues from exports and from various credits.[121] While Turkmenistan may launch a licensing round of 32 blocks in what it considers its portion of the Caspian, there is no queue of Western companies preparing to bid.[122] Any new offshore discoveries of oil and gas are expected to lie in very deep formations of 5,000–7,000 m. If these exhibit the same characteristics of Kazakh offshore deposits with high levels of sulfur and mercaptans, costs will be high.

Turkmenistan's oil production occurs mostly in the west of the country, both onshore and offshore. Of the five PSAs with foreign companies, only three are in production. The Burren Energy and Dragon Oil areas are in western Turkmenistan. Burren is a U.K. company producing about 12,000 b/d from the onshore Nebit Dag area. Dragon Oil, headquartered in Dubai, produces 15,000 b/d from the offshore Cheleken field, and is the only foreign company producing offshore. Petronas and Maersk signed offshore development deals, but neither company has begun production. Panama's Mitro International is a minority partner with Turkmenneft, and their Khazar project is producing in the neighborhood of 4,500 b/d. Burren exports by tanker, through an arrangement with BP, and thereafter by rail to Batumi. Dragon exports in a similar arrangement, as well as swaps with Iran and sends some production by tanker to Makhachkala for further shipment through Russia on the northern pipeline to Novorossiysk. But it is Turkmenneft that accounts for the remaining oil production—by far the majority—mainly for domestic consumption. The attempts of the Russian oil consortium, Zarit, to secure a PSA for oil and gas in the southern part of the Caspian Sea near Iran have so far come to naught, although the press reports negotiations continue.[123]

Natural Gas

It is natural gas, however, which makes Turkmenistan potentially interesting from the point of view of future world gas security. Its gas reserves of

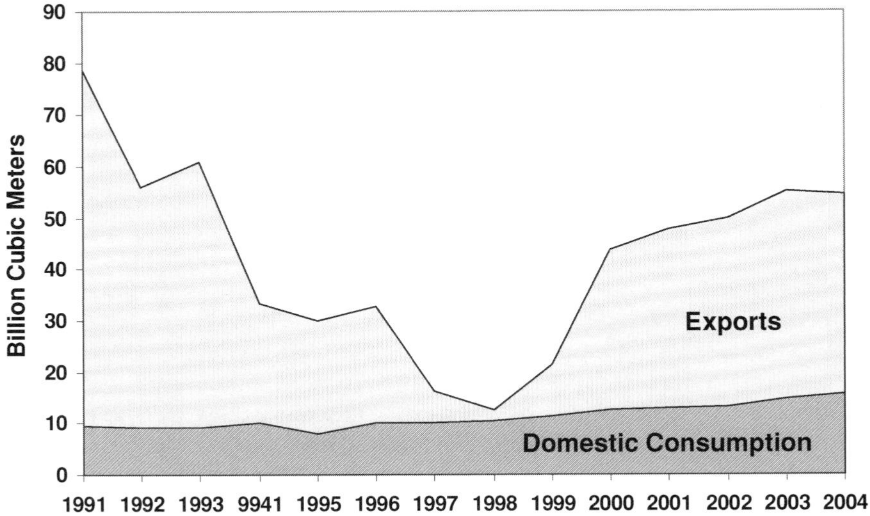

Figure 3.2. Turkmenistan's Gas Production and Exports, 1991–2004
Source: BP, *Statistical Review of World Energy, June 2002 and June 2005.* Reproduced with permission.

2,900 bcm are far larger than its annual consumption of 55 bcm, and nearly all could be exported to international markets were competitive international transportation capacity available. Presently, Turkmenistan remains nearly fully tied to the existing Russian gas pipeline system, except for a small volume of gas exports to Iran. Over the years since independence, gas output has experienced huge fluctuations, and has yet to exceed preindependence levels. Figure 3.2 shows these fluctuations in total gas production and in gas exports since the country's independence.

In 1991 natural gas production exceeded 80 bcm; it fell thereafter to a low of 12.4 bcm in 1998 before rising to 54.6 bcm in 2004. The plunge in production in the 1990s was due to two factors. First, traditional buyers in Ukraine and the Caucasus failed to pay for the gas Turkmenistan shipped. Second, Moscow regarded Turkmen gas as a competitor to sales of its own gas and cut off Turkmen shipments on the existing pipelines to Russia. With no other sales or transportation or outlets, Turkmenistan gas production had nowhere to go, and production declined dramatically.

With revenues to the state severely affected from the decline in gas sales during most of the 1990s, Turkmenistan began to look for alternative gas sales prospects. Beginning in 1997, it sold gas to Iran through the newly constructed pipeline from its Korpedzhe field to Kortkui across the border. Shipments were to grow from 2 bcm/y initially to 7 bcm/y in 2003, and to 12 bcm/y by 2006; the flow has not yet reached the 2003 target. Turkmenistan remains interested in selling to Iran for further deliveries to Turkey, although the robustness of Turkish gas demand will ultimately determine the size of its gas purchases, as also will the state of U.S. policy toward Iran. Sales to Iran, however, could turn into a swap arrangement, with Turkmen deliveries to northern Iran paid back with Iranian deliveries in Turkey. But Iran and Turkmenistan are or may become competitors for sales in Turkey and elsewhere in Western Europe, in which case Tehran may not wish to simply serve as a pass-through route and/or see Turkmen gas sales cut into its own potential sales. Iran could, nonetheless, import Turkmen gas for domestic use in the north, thereby freeing up more domestic gas for export.

After its failed export proposals, Turkmenistan reexamined sales to and through Russia via the existing gas pipelines. It concluded an agreement with Russia in April of 2003, when Gazprom agreed to a 70–80 bcm/y contract for 25 years from 2004 through 2028. Volumes were to build from 6 bcm/y in 2004 to peak levels by 2007.[124] The price was reportedly $44 per thousand cubic meters, or about $1.25 per MMBtu, well below prices of $3.50–4.00 per MMBtu prevailing in Western Europe at the time, and higher than Russia initially was prepared to pay. It should be noted that Russia sold its gas in Europe for about $100–130 per thousand cubic meters, so its Turkmenistan purchases were highly profitable.[125] Turkmenistan was in the first 3 years to be paid half in cash and half in bartered goods and services.[126] At assumed peak volumes of 80 bcm/y, this contract, if fully paid in cash, would earn Turkmenistan more than $3.5 billion annually.

In addition, Turkmenistan at the same time was arranging a 25-year sales agreement with Ukraine, at the same sales price as agreed in the Russian contract. The contract called for the purchase of at least 45 bcm/y from 2007 through 2032. The agreement effectively extended and increased the current 5-year contract ending in 2006, which provided for 34–36 bcm/y, at $44 per thousand cubic meters, paid half in cash and half in bartered goods. At peak levels, if subsequently paid fully in cash at that price, the Ukraine contract would provide Turkmenistan with about $2 billion annually. Figure 3.3 shows the route through Russia to Ukraine.

But in January 2005, Turkmenistan abruptly shut off the gas flow to both Ukraine and Russia in an ongoing dispute over prices. Turkmenistan demanded a price increase to $60 per thousand cubic meters, or about $1.70 per MMBtu, half to be paid in kind. Ukraine's new leadership capitulated and signed a contract to buy 36 bcm/y for 1 year at a price of $58 per thousand cubic meters, half paid in kind, and the gas flow was restored. While agreeing to expand energy cooperation in the future, Russia refused to pay the higher price, arguing that the price was contractually fixed through 2006. Gazprom indicated it could do without 7 bcm of Turkmen gas in 2005 in view of new supplies from independent Russian producers, and proposals from Kazakhstan and Uzbekistan to sell gas. Russia's hard line prevailed; gas flow was restored in mid-April 2005 at the original price of $44 per thousand cubic meters. Moscow agreed, however, to make 100 percent payment in cash rather than the 50 percent in cash and 50 percent in bartered

Figure 3.3. Turkmenistan's Natural Gas Export Route to Ukraine
Source: U.S. Government Map. *Turkmenistan's Natural Gas Export Route to Ukraine*, (INR) 8955, January 2004.

goods.[127] While Niyazov clearly miscalculated the strength of his trading partner, his peremptory demand for higher gas prices sets a precedent, and one should expect that he will repeat his demand for higher prices in the future; he is already suggesting that by the end of 2006 he will demand at least $100 per thousand cubic meters for Turkmen gas. But Turkmen gas has no other place to go, and the country has insufficient funds to build a new pipeline on its own. Russia holds the stronger cards, and Niyazov, in this author's view, was forced to accept Moscow's terms.

Projections for Turkmenistan's future gas production and sales are fraught with uncertainties. These include in the nearer term reserves size, gas quality, pipeline capacity and availability, commitment of the leadership to honoring contracts, and relations with the neighboring countries. For the present, Russia controls the country's gas export routes, with the exception of the line to Iran. For the longer term, Turkmenistan is courting both Pakistan and China as potential natural gas customers. We address each of these in turn.

- There is repeated reference in the trade press that Turkmenistan may have overcommitted sales of its gas reserves in its recent contracts with Russia and Ukraine. Some analysts believe that the country does not have the reserves, the productive capacity, or the pipeline capacity to deliver 80 bcm/y to Russia, 45 bcm/y to Ukraine, and meet as well its commitments to Iran of 7–12 bcm/y. Production in 2004 amounted to about 55 bcm, well below these commitments. The issue of gas reserves adequacy becomes of greater concern if Turkmenistan also moves forward with the resurrected proposal to ship as much as 15–30 bcm/y southward through Afghanistan to Pakistan. Wood Mackenzie certainly questions whether the country has the productive capacity to meet the Ukraine, Russian, and Iran contracts of roughly 140 bcm/y—more than double the current production—for 20 years or longer.[128] That level of exports cumulates over time to 2,800 bcm, yet BP and others estimate the country's gas reserves at 2,000–2,900 bcm. The long-awaited independent certification of proved gas reserves in the large Turkmen Dauletabad field by the respected U.S. firm of DeGolyer and MacNaughton was reportedly released without authorization, and press reports suggested at the time that gas reserves (not defined) amount to 4,500 bcm.[129] Most analysts, however, think this is a gross overestimate in terms of proved reserves available for supporting future production. Wood Mackenzie likewise took a far less optimistic view of future Turkmen gas production. It sees it peak, without a sudden jump in capacity, at 67 bcm/y between 2006 and 2011, and decline thereafter.[130] Exports would be somewhat lower, depending on internal demand, although the latter is expected to remain at a low level of about 10–15 bcm/y.

- In addition, Turkmenistan is also allegedly negotiating with China to build a pipeline and to sell as much as 30 bcm/y to China in the years ahead, presumably linking the new line to the West-to-East Chinese gas pipeline.[131] These negotiations, however, may be more of a means to put pressure on Ukraine and Russia who have been unwilling to pay as much for Turkmen gas as Niyazov has sought. Most experts do not believe that Turkmenistan's gas resources are sufficient to meet sales to Ukraine, Russia, Iran, Pakistan, and China in the volumes stated.
- Turkmen gas quality is also thought to be deteriorating, that is, becoming more sour in terms of sulfur content. The sourness increases the difficulty of maintaining the wells, and raises the costs of removing impurities to provide pipeline-quality gas.[132]
- For sizeable Turkmen volumes to reach Russia and Ukraine in the future, either existing sections of the CAC gas pipeline system must be repaired and rehabilitated, or new lines built. These several pipelines were built some 20–30 years ago to serve Soviet purposes, not those of the now-independent states. Originally capacity was 90 bcm/y, but the lines are presently in bad disrepair, with current capacity estimated at half that amount. One of the CAC lines running from Turkmenistan through Kazakhstan to Russia, however, is being expanded from 5 to 10 bcm/y.[133]
- Complicating Turkmenistan's ability to move its gas in the future is the stated goal of both Kazakhstan and Uzbekistan to increase their gas exports, using the same CAC pipeline system. How much Turkmenistan can or will ship will be determined not only by pipeline capacity but by its neighbors' rights as well. Uzbekistan has announced, for example, that it plans to export some of its gas on the CAC system, and would therefore make only 20 bcm/y of export space available to Turkmenistan.[134]
- There is always a risk that Turkmenistan may not honor its sales contracts, given Niyazov's reputation for unpredictability and arbitrariness in decision making. Normally gas contracts in Western countries contain take-or-pay clauses on the purchaser, so as to ensure a revenue stream to the seller. This revenue stream in turn serves as collateral to secure financing for construction of the pipeline. Yet, in this instance, we do not know whether there exists a parallel produce-or-pay clause that would obligate Turkmenistan as seller to deliver the gas, or else face costly consequences.
- Russia is—and will remain—the key player in determining whether and at what price Central Asian gas reaches markets, at least until substitute pipelines are built that avoid Russia. Thus Russia maintains its monopsony power as buyer, and Gazprom its transportation monopoly, to end markets.

Gazprom and others are looking at the feasibility of conducting a 10-year, $1.3-billion upgrade of the CAC system. In addition, a brand-new

$1.2-billion, 30-bcm/y pipeline through Kazakhstan is under consideration.[135] Yet bringing CAC back up to 90 bcm/y capacity and building the Kazakh line will still not meet the capacity needed for Turkmen gas exports, if the latter approach the contract level of 140 bcm/y. In addition, whether the CAC system is expanded, or a new line built from Turkmenistan through Kazakhstan to Russia and Ukraine, the construction timetables appear overly ambitious and are unlikely to meet the deliveries scheduled for 2007 and thereafter.[136] Both rehabilitation, and/or new construction as proposed by Gazprom, will continue to keep Turkmen gas exports firmly under Russian control.

In the longer term, however, Niyazov is looking at gas sales in directions that avoid Russia. Whether Turkmenistan has enough gas to meet its current commitments is one issue; whether it plans to meet them is another. Niyazov may just be hedging his bets in trying to develop alternative gas markets should he secure better terms elsewhere than those for sales to Russia and Ukraine. These possibilities include the recently resurrected Trans-Afghan pipeline proposal and a gas export line to China.

The Asian Development Bank recently breathed new life into the 15–30-bcm/y trans-Afghan gas pipeline project. The Bank sponsored and completed a feasibility study in 2005 of the proposed 1,680-km line. It found that the project, estimated to cost $3.5 billion, is technically and economically feasible, provided financing is obtained. The project has won U.S. political support, as it passes through neither Russia nor Iran. From the point of view of private investors, interest is not high given the ongoing unrest in Afghanistan. The future demand for Turkmen gas in Pakistan is also uncertain, since Pakistan has substantial domestic gas reserves thought sufficient to cover its anticipated consumption through 2020.[137] While India has been both for and against the project, by mid-2005 it was again interested in it. Most observers think the line is unlikely to be built unless the volumes are sufficiently large to justify the investment, and that requires India's firm commitment. Competing with the trans-Afghan gas line are other proposals from Iran across Pakistan, and a Bangladesh–Myanmar line to India. Complicating all these considerations are Turkmenistan's peremptory demands for higher gas prices from both its Russian and Ukrainian customers, and the as-yet questionable size of Turkmenistan's Dauletabad gas reserves. Financing will be difficult to arrange without sufficient demand, greater political predictability and stability in both Afghanistan and Pakistan, and continued warming in Pakistani–Indian relations. Figure 3.4 shows the alternative routes running from Dauletabad in southern Turkmenistan to Herat and Kandahar, and on to Quetta and Multan in Pakistan. It also shows the northern route, ruled out because of severe winter weather conditions. Other than the substantial political and financial risk, the two key energy questions surrounding this project are whether there is sufficient gas demand to justify the project, and whether Turkmenistan has sufficient capacity to

Figure 3.4. Proposed Gas Pipeline from Turkmenistan through Afghanistan to Pakistan

Source: Adapted from Asian Development Bank, *Technical Assistance for the Feasibility Studies of the Turkmenistan-Afghanistan-Pakistan Natural Gas Pipeline Project*, December 2002, 1 (15, 3). Reproduced with permission from the Asian Development Bank. For more information on development in Asia and the Pacific, see www.adb.org.

honor a new contract for 15–30 bcm/y, in addition to its gas contracts with Russia and Ukraine.

To round out its thinking on future gas exports, the government of Turkmenistan continues to keep alive a proposed gas export project to China. Most observers agree that this project could be possible in the long run, depending on investment in gas capacity and the willingness of China to pay for a remarkably long and expensive pipeline, estimated to cost

$12–23 billion. China's growing appetite for energy may move the project's timetable forward.

The idea of a Turkmen gas pipeline to China first surfaced in 1993, as part of the strategy of reducing Turkmenistan's dependence on Russia for gas transportation. But Kazakhstan also harbors desires to export large volumes of gas to China, and perhaps eventually to Japan. In Turkmenistan's view, the 25–30-bcm/y pipeline would begin near the Uzbek border at Khiva, and link at Shanshan in China in the heart of the natural gas Tarim basin with the Chinese West-to-East gas pipeline. From the perspective of the adequacy of Turkmen natural gas reserves, this export project is subject to the same criticisms made earlier: is there sufficient gas to amortize the project in view of all the commitments to other major export projects? But a pipeline to China could be realized in the next 20 or 30 years, provided that the investment climate improves and sufficient gas reserves are developed to support the project. Figure 3.5 depicts various gas pipeline proposals to serve the Chinese market.

Turkmenistan realizes that it must explore options for non-Russian transportation for its gas exports, although it has yet to succeed in realizing any. Niyazov may have learned a lesson after his most recent defeat by Russia in trying to secure higher gas prices, and by his previous fumbling and delays with the Trans-Caspian project. This project was sidelined by the decline in Turkish gas demand and the decision of Azerbaijan to develop Shah Deniz for export. Turkmenistan most assuredly recognizes the impacts of multiple-pipeline competition as well as the monopsonistic power that Russia currently exerts as the only export system to carry Turkmen gas production. The prospects for gas sales through non-Russian routes, however, are dim at present. Western Europe has many sources of gas other than that which might prospectively move through Turkey, and Pakistan and India have ample gas, but could be supplied—should demand warrant—more cheaply and with less risk by pipeline from Iran or as LNG from Oman. The wish to export to China remains just that, and should Central Asian gas eventually move to China, Turkmenistan will face severe competition from Kazakhstan natural gas also in search of eastern sales.

Conclusion

Conducting business of any kind in Turkmenistan is a high-risk undertaking. The foreign investment climate remains extremely poor. The rules of the game are never fully known, they frequently change, and the rule of law is nonexistent. Given Niyazov's personal control over all aspects of the country's economic and political life, foreign firms are reluctant to invest. Only a few small foreign companies are presently active, producing small quantities of oil and no natural gas. Noticeably absent are the large

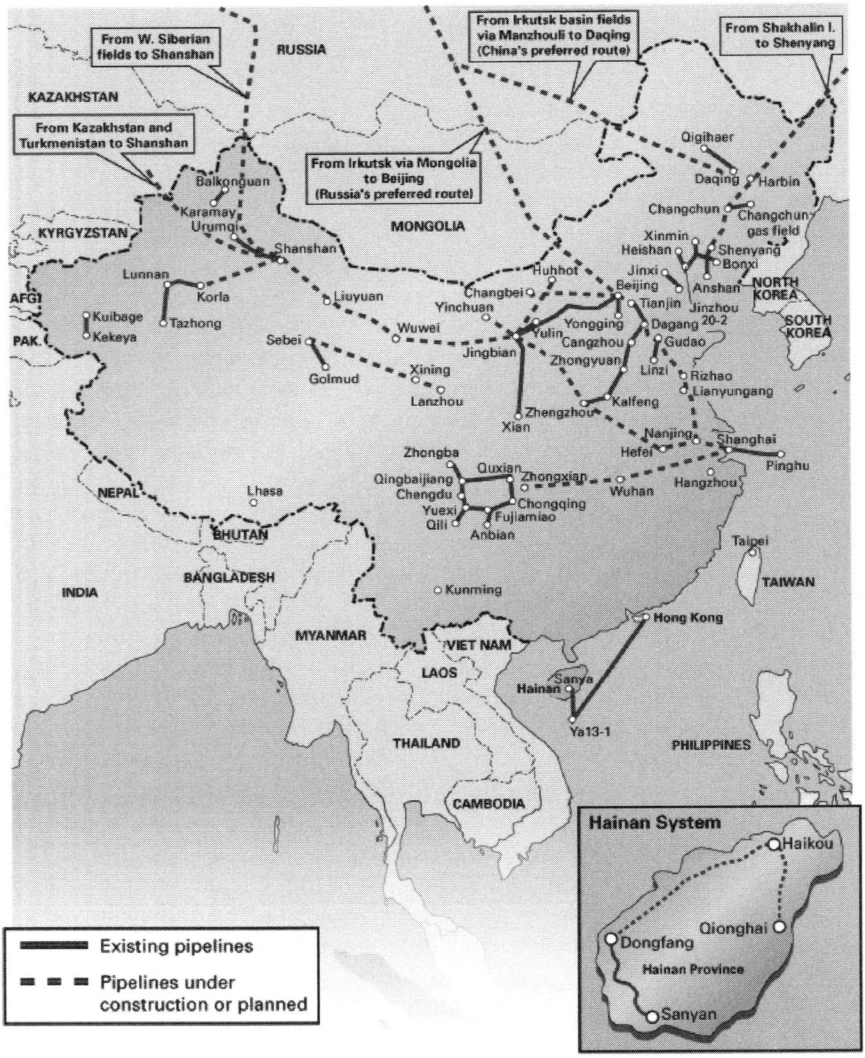

Figure 3.5. Possibilities for Central Asian and Russian Gas in the Chinese Market
Source: Anonymous, "China launches massive gas pipeline campaign," *Oil & Gas Journal*, Vol. 98, Iss. 23, June 5, 2000: 65. Reproduced with permission.

international majors. As a representative of one of the latter told the author, "the politics are simply not right."

Turkmenistan's politics cast a great shadow on the country's prospects for energy development. The world has more than enough gas to do without gas from Turkmenistan. Russia, however, would suffer, since it has become used to exerting its monopsony power to buy Turkmen gas at below market prices to meet Russian internal demand. Iran could do the same. Without Turkmen gas, both would be forced to develop new deposits at considerably higher costs. If and when the Turkmen investment climate improves, the government provides stability and predictability under the rule of law, and anticipated demand merits, international investors will display interest in the country's energy resources and can be expected to risk capital to develop them.

IRAN

Iran believes it has important national interests in the Caspian region, as much from history and geography as from any identified hydrocarbon resources in its sector of the Sea. Until the dissolution of the Soviet Union, it and the USSR ruled the Caspian Sea and its environs. With the dissolution of the Soviet Union, Russia assumed all rights and responsibilities of formerly Soviet international treaties. Today the two countries are rivals in energy development, as well as in regional politics.

Iran's 740-km coastline runs from the Azerbaijan border in the west to the Turkmen border in the east. Nearly a quarter of its population is Azeri (Azerbaijan historically formed a part of Iran), and relations between the two countries have not always been amicable. Iran fears (perhaps with good reason) that its Azeri provinces may be secessionist-prone, while Baku views Iran as a supporter of Armenia, with whom Azerbaijan is still technically at war.

Iran's 132 billion barrels of proven oil reserves place it second to Saudi Arabia in world oil reserves.[138] Its natural gas reserves are estimated at about 27,500 bcm, second in size only to those of Russia. Both these figures include the two countries' Caspian reserves, which are currently estimated at zero or negligible. With most producing or identified oil fields located in the south and west of the country, Iran exports from terminals in the Persian Gulf. It increasingly engages in Caspian oil swaps, imports a modest amount of gas from Turkmenistan, and exports gas to Turkey. So far its exploratory efforts in the Caspian have been minor, with no commercial discoveries. Large new onshore oil fields may prove far more promising than those that might be found in the Caspian, in terms of competing for development capital, as are the proposed LNG export facilities from Iran's Persian Gulf ports to sell gas to world markets. Iran has long played the spoiler for gas and oil exploration and development of its share of the Caspian; it wants to maintain its claim to a greater share of the seabed than it secures by median line division. In 2005, it changed its approach, however, and now appears prepared to invest relatively small amounts of capital for exploration of prospects in its (not-in-dispute) share of the Sea. The bulk of its development capital seems firmly committed to oil and gas projects in the south, particularly the South Pars gas field.

In terms of transportation, the least costly route for moving Caspian oil to world markets on ocean-going tankers would be by pipeline south through Iran (such as by the proposed Kazakhstan–Turkmenistan–Iran line), or through a sizeable system of swaps by which Caspian oil is received in the north in exchange for equivalent barrels at Iran's southern terminals. Yet from an energy security perspective, regardless of the continuation of U.S. sanctions, such pipeline deliveries or swapped barrels allegedly heighten the concerns of consuming countries about becoming overly dependent on energy exports from the Persian Gulf and the Middle East, particularly through the Straits of Hormuz.[139] What this line of reasoning fails to note, however, is that oil is a fungible product, with prices adjusted for quality, and there is a world oil price, regardless of where the oil originates. Any events that are perceived to raise risks or costs, be they political/military tensions, or actual supply interruptions, will rapidly affect oil prices worldwide, not just the price of oil from the Persian Gulf and the Middle East.

Iranian oil production in recent years has hovered around 3.5–3.9 mmb/d, with net exports of about 2.5 mmb/d. Current production and sustainable capacity are about 4.1 mmb/d, slightly over Iran's OPEC production quota level.[140] The country never regained its prerevolutionary production levels of more than 6 mmb/d in 1974. Damage from the 1980s war with Iraq, poor maintenance and overproduction, absence of technical capabilities, and U.S. sanctions have all depressed Iranian production levels. In the 1990s, the goal of restoring production to 5 mmb/d was repeatedly announced, but not realized. Iran has announced ambitious plans to raise output to—depending on the source—5 mmb/d by 2005, and to 7 mmb/d by 2015, or alternatively the former target by 2009 and the latter by 2024.[141] It missed the 2005 target, producing only 3.9 mmb/d that year. One of the problems the country must overcome is that mature fields' output is declining at an estimated 300,000 b/d, and higher gross production targets must more than offset this decline.[142] With sufficient investment in development and maintenance, and the proposed development of the sizeable Azadegan, Bangestan, and other new fields, these production goals may be possible in the years ahead, although such expansion will contravene Tehran's commitment to OPEC in terms of limiting production. The key to expansion will be the amount and source of funds. Without changes in the country's ability to attract foreign investment, and/or political changes in its relationships with Western countries, both the expansion and its timetable are unlikely to be realized.

Domestic Political Turmoil

For the 8 years of President Mohammad Khatami's administration, gridlock between elected reformers and unelected conservatives characterized Iranian politics. It was a gridlock of chaos between those with democratic and those with clerical credentials. The country's political divisions diverted Iran's attention from Caspian oil and gas developments, save for Tehran's

Mahmoud Ahmadinejad [AP Photo/Julie Jacobson]

Iran: A Newcomer in a Conservative Landslide

Mahmoud Ahmadinejad was elected president of Iran in June 2005, defeating ex-President Akbar Hashemi Rafsanjani in a surprise landslide run-off vote. The result put the reins of power firmly in the hands of religious conservatives who now control all of Iran's elected and appointed institutions. He is as yet an unknown identity on most issues of both domestic and foreign policy, although it is well known that he has no intentions of curtailing Iran's nuclear program in the face of Western pressure, and has in fact resumed uranium reprocessing.[143] While he holds office for 4 years, it is too early in his term to take his full measure. It appears, however, that unless U.S. or Iranian policy changes, improvements in the bilateral relationship are slim.

In the oil and gas sector, the new government's target level is about 50 percent higher than the current production of approximately 4 mmb/d. The date for reaching that target is uncertain. It is also not clear whether that target will involve expedited exploration and possible development in Iran's undisputed share of the Caspian Sea. Given that the country continues under U.S. sanctions, meeting this target will require substantial expenditures on equipment and technology from companies and countries willing to risk sanctions. Iran denied that it would curtail oil production in the event the IAEA board voted to report Iran to the UN Security Council for its nuclear activities. With India's vote in favor of the IAEA resolution, Iran may stall on moving forward with proposals to deliver pipeline gas or LNG to India. In future UN Security Council deliberations, Russia and China are not likely to support the U.S. in favoring sanctions on Iran for its lack of cooperation on nuclear matters. Should they alter their positions, however, such that the Council ultimately imposes sanctions, Iran's response, as well as that of the West, is uncertain. Iran has threatened to withdraw from the Nuclear Non-Proliferation Treaty and continue to pursue what it sees as its right to engage in uranium reprocessing for the development of peaceful nuclear power. Western countries face a dilemma as to what additional means, diplomatic or military, might lead Iran to change its mind.

Born in 1956, Ahmadinejad has lived the majority of his life in Tehran. He earned a PhD in transportation engineering from the Iran University of Science and Technology, and thereafter became a lecturer at that university. He has categorically denied his involvement in the 1979 hostage taking of Americans at the U.S. Embassy in Tehran. He saw military service in the Iran–Iraq war beginning in 1980 and became a member of Iran's Revolutionary Guards in 1986. Following the war, he entered regional politics, and also advised the national minister of culture and Islamic guidance. After a term as governor of Ardabil province, he was appointed mayor of Tehran in 2003. He appears to be a skillful politician, able to use his humble roots as a son of a blacksmith to demonstrate his position as a simple man of the people. His impeccable Islamist credentials won him support among conservatives and Islamic clerics, while his populist promises to end corruption and to provide jobs and raise wages attracted voters from Iran's poorer population. The Tehran business community was in general not enthusiastic over his victory.

---

insistence on a bigger share of the Caspian seabed than other littoral states were prepared to recognize. The country elected a new president, Mahmoud

Ahmadinejad, in mid-2005, and it will take time to discover his administration's policies toward Caspian issues. He is widely viewed as a conservative and religious hardliner, and has called for Israel to be wiped off the map. His course is as yet unknown in terms of what specific directions his domestic economic, energy, or foreign policy will take. By resuming uranium reprocessing, ignoring the IAEA's requests for inspection of nuclear facilities, and threatening to withdraw from the Nuclear Non-Proliferation Treaty (NPT), he has taken a tough line with the United States and the West. The United States is likely to experience growing difficulty in dealing with Iran on nuclear issues and on Iran's alleged support for international terrorism.[144] The country's nuclear program is a source of enormous prestige to its citizens and is one issue of agreement between factions that disagree on other issues. But it is clear that after the fractious past of reformists confronting conservatives, current power is now firmly in the hands of conservatives, and chances of gridlock in the future may be less than in the past.

The United States has steadfastly maintained that Iran supports nuclear weapons development, while Iran has long rejected any curbs on its nuclear program and stated its intent to develop its nuclear capacities for peaceful use. Given Tehran's extensive hydrocarbon resources, many question why Iran should pursue peaceful nuclear power, if that indeed is its sole goal. The answer is clear, however, for the more that hydrocarbon fuels can be freed from use in domestic electric generating plants, the greater the likelihood for increased world oil and gas exports and revenues. In addition, the country's rapid population growth signals rising internal electricity demand, to be met by both fossil and nuclear generating plants. Iran intends to develop new fields for export of both oil and gas, particularly in the south and west. The oil and gas reserves in this region are far greater and already developed than those likely to be found in Iran's Caspian Sea area.

Standing Firm on Seabed Delimitation

Despite its inward focus, however, Iran is participating once again in meetings with its neighbors to resolve the issue of division of the Caspian Sea. Prior to the fall of the Soviet Union, Iran and the USSR shared the Sea under the treaties of 1921 and 1940. These treaties did not cover mineral resources, but did address conditions for fishing and navigation. Since the creation of new littoral states out of the old USSR, however, Iran has been unable to reach agreement with its neighbors on the appropriate division of the Sea for use of either the water column or of the seabed resources. For 15 years the littoral states have met to address seabed delimitation, but agreement remains elusive. Iran continues to hold out for a 20 percent equal share of Sea ownership, whereas division by the modified median line method—accepted in 2003 by Kazakhstan, Russia, and Azerbaijan—would give Tehran only 12–13 percent of the seabed. Iran was quick to declare illegal the agreement accepted by its three neighbors. We believe that Iran is playing the role of a spoiler, trying to slow down Caspian development of gas and oil resources, even though it has little intention of committing major resources to exploitation of the area, given its extensive development plans in the south. Iran continues to hold out for Sea division according to the principles of "justice, fairness, equality and equity," which are generally interpreted by others to mean nothing less than 20 percent seabed.[145]

Until agreement on seabed division is reached, no realistic estimate of Iran's Caspian resources can be made. With no current Iranian drilling activity or production in the southern portion of the Sea, it is pure speculation as to what the area might hold were drilling to occur. The few exploration wells drilled in the past yielded no commercial discoveries.[146] Current reserves estimates, however, for oil and gas in the areas near/in the Iranian portion of the Sea are estimated at close to zero or listed as not available. Many energy company representatives consider the Iranian sector of the Sea, whether 20 percent or 12–13 percent, to be devoid of any commercial deposits of oil, although deep gas may be present.[147] The southern sector of the Sea, in both Azeri and Iranian areas, is thought to be gas-prone rather than rich in oil. In part this view stems from the belief that the Azeri sector south of the Apsheron trend is a gas-prone one. Iranian officials, nonetheless, say there may be as much as 10 billion barrels of oil in their sector, but this may be wishful thinking since nothing commercial has been found to date.[148] Absence of Iranian exploratory activity, however, does not equal absence of interest in the area. Iran showed its commitment to its general area of the Sea in a widely publicized incident in July 2001, when its military forces drove off an exploration vessel accused of operating in waters that both it and Azerbaijan claim.

Of interest is why Iran pursued a nondevelopment policy and chose not to come to agreement on seabed delimitation. By steadfastly holding out for a larger share of the Sea, it delayed the potential revenue stream from any future discoveries and production in "its sector," however defined. It no doubt views a larger portion of the seabed as giving it more reserves potential than a smaller share, but this potential gain must be assessed against the certain delay—and thus losses—of realizing potential revenues. The issue is how long can Iran hold out and forgo potential revenues. The worst of all possible outcomes for Iran is that it abandons its current position and settles for division by median line method, losing not only the share of the seabed issue, but also the opportunity to have generated potential revenues earlier in time had it agreed to the smaller sector. If this occurs, Iran has lost doubly, for it could have reached such an agreement years ago and begun at that time an exploration and development program. In a study addressing these issues, Hossein Askari concludes, using a discount rate of 10 percent and an oil prices of $15 per barrel, that if it takes more than 5 years for

Iran to achieve its desired share of the Sea, it should settle for the lesser share.[149] Askari also notes, however, that the 5-year horizon keeps slipping with the passage of each year; continued delays diminish the discounted future revenues.

Considerations along the above lines may have influenced Tehran's announcement in 2003 that it would begin within the next 2–3 years to develop its portion of the seabed that is not in dispute.[150] Tehran apparently now recognizes that competition from the other four littoral states is forcing it to abandon its nondevelopment policy prior to agreement on sea division by all five. According to the trade press, Iran realizes it is being left behind in Caspian activity, and plans to proceed with exploration, and development if the latter is cost-effective.[151] It announced that it would start drilling in the Caspian offshore in 2005–2006 from a rig under construction at Neka, but at the time of this writing drilling had not begun. Recent reports suggest that Iran wants to attract Russian and other international companies to bid for drilling and production rights in the undisputed Iranian sector of the Sea.[152] While seemingly abandoning the nondevelopment policy, Iran continues to hold firm on the delimitation issue.

Tehran to date has lost out not only in Caspian production, but also in transport. It pushed hard to be selected for the main export line for Azeri oil from the ACG fields, but U.S. sanctions policy was a major reason that the Baku–Tbilisi–Ceyhan route was selected. Most observers of the route selection process noted that the Iranian alternative to ports on the Persian Gulf would have been the cheapest pipeline choice, with onward tanker transport to the energy growth regions of Asia.[153]

Swaps

In the meantime, Iran is carving out a different role, both as a user of Caspian and/or Russian oil, and as a transit nation offering a far cheaper alternative to oil shipments on the BTC pipeline.[154] It is striving to become a key hub for oil received at its northern ports to be swapped for Iranian oil exports in the south, aimed primarily at growing Asian markets. Iran is also looking at rail transport from Caspian locations to its southern ports. Swaps are attractive since Tehran's population and demand centers are in the north. Swaps costing $1.50–2 per barrel reduce Tehran's transport costs to feed the northern refineries at Tehran and Tabriz. The funds earned are used to reconfigure the refineries to process high-sulfur Caspian crudes.[155] Swaps are also attractive to Caspian and Russian producers, since swaps reduce costs below those on pipelines to the Black Sea. Swaps have at times averaged from 110,000 to 150,000 b/d, but capacity could rise to 700,000 b/d to 1 mmb/d, particularly if the Russians agree to swap up to 300,000 b/d.[156] Swaps, however, suffer from short-run sweet versus sour crude price disparities between the Asian and European markets. When European prices for sweet crude are relatively high, Russian and Caspian crude producers seek sale in that market and are less interested in swaps through Iran to Asian markets. Since Kazakhstani oil tends to congeal at low temperatures, swaps with Iran diminish in the winters (especially since Iranian refineries are geared to higher-quality oils).

- In mid-2005, Kazakhstan and Turkmenistan together swapped about 60,000 b/d with Iran.[157] In early 2006 total swaps at Neka were running at 80,000 b/d and expected to rise to 100,000 b/d.[158] KazMunaiGaz is building two oil storage terminals with a capacity of 150,000 barrels in northern Iran, which when completed will permit Kazakh oil more flexibility in the timing and volume of swaps.[159]
- Russia's LUKoil began swaps with Iran of 25,000 b/d in 2003, rising to 40,000–50,000 b/d in 2004. In 2005, reports indicated that LUKoil could deliver 120,000 b/d to Neka for swap.[160] Russia continues to face severe capacity constraints on its export pipelines, and swaps offer it a convenient method of moving additional barrels.

As overall Caspian oil production mounts, swap arrangements with Iran could also rise, or be marginalized by other transport opportunities. The inauguration of BTC in 2005, together with any future use of proposed Bosporus bypass routes and the realization of the oil line to China, may prove more economic and lead to the desertion of barrels presently swapped for pipeline movement to other destinations. The United States is on record as opposed to swaps with Iran, but Washington is unlikely to do anything about them. In the South Pars case, the White House/State Department decided to look the other way because key strategic allies, such as Russia and Germany, were involved. Similarly, Kazakhstan is a key country if only because of the amounts U.S. majors have invested. If oil swaps through Iran become sufficiently large in the future to justify a north–south underwater Caspian Sea pipeline, the United States may take greater notice.

Natural Gas

In natural gas production and transport, swaps are not part of the Iranian energy landscape. Iran is looking both westward and eastward to market its extensive reserves of natural gas. Even if the southern Caspian Sea area, under a future identified and settled Iranian claim, turns out to be gas-prone, Iran has plenty of unmarketed gas for sale already, and does not need additional Caspian supplies. For more than 20 years, the country has proposed gas exports by pipeline through Turkey to Europe. Its key market threat is that Azerbaijan, followed by Kazakhstan or Turkmenistan, might successfully compete for a slice of the European market, thereby meeting European incremental demand and depriving Iran of future European sales.

Looking to the east, Iran continues to try to win Pakistan and India as firm customers for a proposed gas pipeline linking the three countries. India and Iran in 2005 signed an $18-billion, 25-year gas contract, with deliveries to begin in 2009 or 2010 at roughly 4 bcm/y, building to more than 36 bcm/y by 2013–2014.[161] The gas will originate from Iran's South Pars gas field. These two countries and Pakistan also signed a memorandum of cooperation for constructing a $6-billion, 2,735-km gas pipeline running from southwestern Iran to Pakistan and ultimately to New Delhi. For the near term both Pakistan and India have adequate domestic gas, but their rapid growth suggests that they will need additional supplies by 2020. The memorandum suggests that Pakistan will take up to 1.5 bcm/y if the pipeline moves ahead. For other future Pakistan and/or Indian gas sales, Oman and Qatar are potential competing suppliers by pipeline or with LNG. Warming relations between India and Pakistan underlie the memorandum, as do economic considerations: were Pakistan to interrupt future deliveries to India, the latter expects to have alternative sources from either Oman or Qatar, or perhaps ultimately from Turkmenistan.

It seems clear that Iran in the near term has stolen a march on Turkmenistan by selling large volumes of gas to the subcontinent. Whether the Iran–India–Pakistan agreement puts at risk the realization of the proposed Trans-Afghan Pipeline designed for delivering 30 bcm/y of Turkmen gas to both India and Pakistan remains to be seen. The Iranian proposed route will serve the Pakistani city of Multan, which was one of the cities proposed to be fed by the southern alternative of the Trans-Afghan line. Given India's strategy of rapid gasification in order to decrease its oil consumption, there are those who believe that ultimately New Delhi will show a greater level of interest in Turkmen gas than it has to date. Should the Iranian gas pipeline be completed to New Delhi, there remains the possibility that it could be further extended to serve China, thus competing with possible Turkmen and Russian Kovytka gas supplies seeking Chinese markets.

Iran began pipeline deliveries of gas to Turkey in 2002; these are slated to build to 10 bcm/y by 2007, but there are questions, as noted previously, as to how much Iranian gas Turkey will actually buy, even under its take-or-pay contract, given its overcontracting for gas from Russia, Iran, and Azerbaijan. DOE/EIA estimates that Turkey has already bought 25 percent more gas than its anticipated demand in 2010.[162] According to one Turkish energy expert,

> The paradox is that Turkey is trying to get out of its current energy mess, which is the direct result of its overpurchase of Russian gas, through the possible re-export of Iranian gas in order to be able to purchase additional Azeri gas.[163]

With Greece and Turkey agreeing to connect their gas pipeline grids, gas sellers including Azerbaijan, Iran, and Turkmenistan hope to sell in Europe. All will face keen competition there from Russia, Algeria, Norway, the Netherlands, and with respect to LNG as well. Because of the expense of long-distance gas transport, Iran and Caspian sellers may need to displace existing long-term Russian gas sales in the Balkans and Central Europe by convincing Gazprom to engage in gas swaps, thus freeing up Russian gas to meet the demand in France and Germany, for example, while Caspian gas serves the Balkans and Central Europe. In another approach to ultimately selling gas in Europe, Iran in 2004 signed an agreement to sell gas to Armenia, a project under discussion since 1992. While Iran buys relatively small quantities of Turkmen gas to supply its power stations in the north, it will be reluctant to serve as a pass-through to Europe for additional—and competing—Turkmen supplies.

Iran's gas connection to Armenia could be the first step in a line that runs to Supsa in Georgia, underwater to Feodosiya in Ukraine, and then onward to Europe. Given the difficulties in moving its gas to and through Turkey, Tehran is, through Armenia, trying to open the door to more distant, Western markets. The agreed price for sale to Armenia is $84 per thousand cubic meters (approximately $2.40 per MMBtu), about twice what Uzbekistan and Turkmenistan are receiving for their gas sales to Russia. Armenia will pay in part with electricity sales to Iran. The Azeri exclave of Nakhichevan may also be connected to the gas line. Armenia, however, would take only some 0.7–1.5 bcm/y, and major capacity additions would be necessary should Ukraine buy 10 bcm/y and Europe 50 bcm/y. Turkmenistan also is eager to have this potential alternative outlet for its gas, despite Iranian reluctance to serve as a transit route. While the EBRD is prepared to finance the project through Armenia to Western Europe, major uncertainties remain concerning the proposed sizeable and costly extensions to Ukraine and beyond. These possible extensions are also a concern for Moscow, which sees Iranian and Central Asian gas as serious future competitors to it in both Ukraine and Western Europe. In fact, a Moscow newspaper considered the Armenian agreement with Iran as an "anti-Russian gas project."[164]

Politics and Sanctions

How the politics of U.S. sanctions against Iran will play out is not clear. Although Tehran strongly condemned the events of September 11, 2001, the United States' identification of Iran as part of the "axis of evil" in January of 2002 certainly nipped any nascent warming of relations between the two countries. As of December 2003, some of the sanctions were temporarily suspended following the severe earthquake in the city of Bam; the suspensions were extended in April 2004 for another 90 days as a humanitarian and goodwill gesture. Further easing of U.S. sanctions appears unlikely and any steps toward rapprochement will be heavily influenced by Iran's future behavior on nuclear issues, as well as its stand on international terrorism and human rights.

The United States has not repealed ILSA, and full repeal in the near term looks unlikely. The Act comes up for renewal in 2006. Libya, however, is now open for energy development, and diplomatic relations between it and the United States have resumed. The sanctions policy has been remarkably ineffective in changing the behavior of Iran's leadership, and it appears to be fraying at least in the energy field, as various exceptions are permitted.[165] Economic literature suggests that ineffectiveness is normally the result when unilateral sanctions are not widely supported by other countries, and thus it is no surprise that Iran has not been brought to heel. Nonetheless, ILSA certainly gives the conservative Iranian government leadership a scapegoat for its own economic mismanagement. In addition, the Act imposes secondary penalties on non-American companies doing business with Iran, but these have been frequently overlooked:

- After 3 years of bargaining with Iran, Japanese companies in early 2004 agreed to move ahead with the $2.6-billion development program in the southern portion of the large new Azadegan field located near the border with Iraq. But U.S. pressure on Japan led to the withdrawal of Tomen Corp., while Japex as a former Japanese state oil company has been effectively dissolved, leaving only Inpex. Inpex now holds 75 percent of the project, with Iran's NICO holding the remainder, under a buyback contractual arrangement, generally disliked by participant companies.[166]
- Washington did not challenge Turkey's gas purchase agreement with Iran, as it views Turkey as a strategic ally and chose to look the other way.
- Total's agreement to develop portions of Iran's Sirri field, previously offered to ConocoPhillips, was not challenged under U.S. sanctions. Nor was Total's agreement with Gazprom and Petronas to develop Iran's South Pars gas project challenged. Rather, the United States viewed South Pars as a "special case" and granted Total a waiver of sanctions in exchange for French pressure on Iran to stop its support of terrorism.

Upstream exploration and production work in Iran is moving ahead, despite ILSA, but the United States remains adamantly opposed to any new international pipelines through Iran. The next test of this view will come if and when a new proposal emerges, perhaps offered by Kazakhstan, to build either an oil or gas line southward through Iran. Should the Kazakhstan–Turkmenistan–Iran oil pipeline proposal take on new life, its supporters will have to confront and overcome this international political hurdle.

Conclusion

U.S. companies would jump at the chance to return to onshore, non-Caspian Iran were sanctions to be lifted, even though Iran's buyback contracts are less attractive than PSAs elsewhere.[167] Nonetheless, the politics, fiscal regime, and institutional arrangements would have to be predictable to ensure foreign interest, whatever the anticipated returns. The reserves are vast, they are of far lower cost to develop than those of the Caspian, the country is not land-locked, and a transport infrastructure is largely in place. If we look some years ahead, it is possible to envision a time when Iran is no longer a pariah or part of the axis of evil, relations with the United States have been repaired and restored, and an oil or gas pipeline through or serving Iran is no longer a U.S. policy concern.

For such a future, several changes in Iran's behavior and policy would have to occur, according to the current U.S. position, including:

- Termination of its nuclear program and its efforts to acquire nuclear weapons and other WMD;
- Termination of its involvement and support for international terrorism;
- Termination of its opposition to peace efforts in the Middle East; and
- Improvements in its human rights policies and practices.[168]

For such a future as well, the United States' behavior and policy would have to change as suggested in the Council on Foreign Relations 2004 study entitled "Iran: Time for a New Approach."[169] The Council notes that the current policy toward Iran has no tangible benefits, and that the United States should rethink its negative approach and begin to reach out to Iran to pursue constructive engagement. We thoroughly concur; it is blindingly obvious that the policies of successive U.S. administrations have failed with respect to Iran. In the meantime, in a repeat of events in Libya when that country was off limits to U.S. energy companies, the forced absence of U.S. firms from the Iranian oil and gas sector has allowed non-U.S. firms to sign much promising Iranian acreage for future development.

RUSSIA

Russia's energy sector has led the country's remarkable economic growth in recent years. Given its substantial resource endowment, Russia is a natural exporter of both oil and gas to neighboring China to meet the latter's burgeoning demand. Both countries would benefit from future energy trade. Russia could conceivably use its current and expanded transportation system to carry Central Asian supplies to China as well. But Russia's activities in its Caspian region have been minimal to date, not altogether surprising since its resources are so vast elsewhere.

Russia's production and exports have soared and added to world oil supplies in recent years. Growth in oil production in 2004 amounted to about 740,000 b/d, followed by additional growth in 2005 of some 26,000 b/d, and was almost all for export in both years. The United States and other consuming countries look to Russia as well as to the newly independent Caspian countries as an alternative to Middle Eastern supplies. The country's notable economic improvement and energy production expansion since the 1998 debt default and ruble devaluation attracted foreign investor interest, most notably of BP in its early-2003 venture signed with the Russian company TNK. But the Russian rules of the game since that time have become more onerous on both domestic and foreign energy companies, and only a few new ventures have been signed. These include minority interests in Russian ventures purchased by ConocoPhillips and Total. President Putin has systematically moved to tighten state control over the Russian energy sector. With the election of a compliant parliament in 2003 and his own resounding reelection in 2004, Putin gained the political power to impose higher taxes and fees, to rescind licenses, to favor certain energy companies over others, and to alter the terms of doing business.[170]

Russia continues in both political and economic transition. Its moves away from the democratic path toward more central and authoritarian leadership are well documented in Michael McFaul, Nikolai Petrov, and Andrei Ryabov's 2004 book *Between Dictatorship and Democracy: Russian Post-Communist Political Reform*.[171] The Soviet empire of old, together with its command economy, are gone and will not come back, but the country's economic future as well as its democratic progress remain uncertain. After the chaos of the Yeltsin years, most Russians wanted law and order and consistency, and favored the return of power to the president to provide the stability they desired. This centralization of power in the presidency defines the government's dealings with the oil industry, and these often appear unpredictable and arbitrary.

Russia's reserves of both oil and gas are far larger than those of the other Caspian littoral states, and the country is by the size of its production and exports a major player on international oil and gas markets. Its oil reserves are variously estimated as 60–72.3 billion barrels, or from 4.6 to 6.6 percent of the world's total. With more than 27 percent of the world's natural gas, it has estimated reserves of 48,000 bcm. Its production of both oil and gas has risen sharply in the last several years. Oil output rose 50 percent from 6.2 mmb/d in 1999 to 9.5 mmb/d in 2005, and gas production rose slightly from 551 to 589 bcm/y in 2004. In 2005 Russia exported 6.8 mmb/d of crude and products.[172] Russia's announced goal is to increase oil and gas production by roughly 50 percent by 2020.[173] Plans on the books call for nearly 12 mmb/d of production by 2012, provided oil prices stay firm; some experts believe Russia could produce 12 mmb/d now, if it had the appropriate export capacity. Depending on the second West Siberian

Yamal gas project's progress, the goal of lifting gas production from about 590 bcm/y in 2004 to 900 bcm/y by 2020 may take somewhat longer.[174]

Russia as a Caspian Energy Producer and Transporter

Very little of Russia's extensive oil and gas reserves, however, appears to lie in the Caspian region. DOE/EIA shows no proven gas reserves in the Russian Caspian Sea sector, and estimates proven oil reserves there at only 0.3 billion barrels.[175] Indeed, as long as it has long-term access to cheap Turkmen gas, Russia is not likely to be concerned with any potential Caspian gas reserves in its own sector of the Sea. With the agreement between Russia and Kazakhstan on the boundary of the seabed resources, Russia's hopes for a piece of Caspian *production revenues* appear to lie largely with future joint development of oil fields lying on its median line border with Kazakhstan. While there are other untapped structures lying in the Russian sector of the sea, exploratory drilling has yet to commence. The Russian government estimates total reserves, not proved reserves, in these fields at about 385 million barrels of oil, and more than 200 bcm of gas.[176] There is no certainty that these fields will be commercial to develop; until exploratory drilling takes place, reserves estimates are always speculative. The border-straddling fields appear to contain mostly natural gas, and development has been postponed or deferred, due in large part to higher Kazakh tax takes together with captive Turkmen gas supplies.

Development of the fields on the median line with Kazakhstan or of the fields wholly in the Russian sector depends on a number of considerations. These include not only the tax regimes but also the total capital required and the fraction to be provided by the state-owned firms (Rosneft and KazMunaiGaz, both chronically short of funds), the fraction of natural gas contained in the oil, and the availability of equipment. KazMunaiGaz may also prefer to devote its scarce capital resources to those fields that lie fully within Kazakh borders, rather than sharing projects with Russia. Should the fields be rich in gas rather than oil, they are likely to be developed for sale to Russian export markets, given the absence of major gas markets for Kazakhstan.

Russia's interest in the Caspian may be as much tied to its hopes for *transportation revenues* from oil and gas pipeline systems as it is to revenues from potential Caspian production:

- The tariff to move oil on the CPC to Novorossiysk is about $4.00 per barrel, although it may rise to $5.20 per barrel if Russia prevails in its move to raise tariffs, despite the objections of the other owners of CPC.[177] At the lower rate and initial capacity of some 545,000 b/d, this generates $2.2 million per day or $800 million annually, the majority of which accrues to Russia. When the line approaches its full capacity of

1.3 mmb/d, transit charges at the lower rate will amount to $5.2 million per day or nearly $1.9 billion annually. With the addition of pump stations or DRAs, throughput could be increased to close to 2 mmb/d, which would result in commensurately higher transit revenues. Once the line has recovered its capital expenses, tariffs are expected to fall.

- Russia and Azerbaijan negotiated transport terms for Azeri oil to move on the northern route to Novorossiysk. Russia sought a 15-year agreement but Azerbaijan wanted lower transit charges and only a 1-year agreement, presumably with an eye on the BTC pipeline becoming operational in 2006.[178] The northern route moved some 54,000 b/d of Azeri oil in 2003; at a tariff of $2.14 per barrel, these shipments earned Russia more than $42 million in 2003.[179] By contrast, the tariff on the western line to Supsa is only $0.43 per barrel, giving the Azeris ammunition to push for a reduction in the tariff on the Russian route.[180] According to press reports, Russia offered lower tariffs to keep the Azeris' barrels flowing. The Russians have long maintained that the northern route's capacity could be increased to 300,000 b/d.[181] With AIOC's announcement that it will continue to use the western line but not the northern one after BTC begins deliveries to Ceyhan, the northern line will for the most part lie idle or serve as a backup.

With rising production, Russia has been strapped to find ways to move its oil to world markets. Transport capacity is the constraint, not productive capacity. Russia is improving sea terminals and expanding pipelines serving them. At the same time, some internal systems are underutilized, like the northern route. Russian rail and barge shipments of oil have increased in recent years. Owing to lack of pipeline export capacity, rail shipments now amount to 1 mmb/d, even though they are far more costly than shipments by pipeline.[182] Russia's interest in Caspian-region transport to the south has been focused on swaps from Astrahkan to Neka.

Russia's relation to the Caspian area is totally different when it comes to natural gas. While it has ambitious plans to develop additional West Siberian Yamal gas and move it southwest to both internal and international customers, Russia realizes that field development will be costly. As a result, it has looked to Turkmenistan, Uzbekistan, and Kazakhstan as relatively low-cost sellers of gas (and who have virtually no alternative outlets to existing Russian gas pipelines) to bridge the supply gap until Yamal starts delivering sufficient quantities in the future. Gazprom buys from Central Asia at low prices to meet internal demand, which is as noted previously still price-controlled, and sells gas internationally at world market prices. This strategy defers outlays for large capital expenditures on Yamal and increases the current revenue stream. This two-tiered gas pricing system, however, will change if and when Russia becomes a member of the WTO, perhaps in 2006, and thereby agrees to WTO rules.

Russian Companies in the Caspian Region

The major Russian players in the country's oil and gas business are the state companies Gazprom, Rosneft, and Transneft, along with the private companies Yukos, LUKoil, Sibneft, Surgutneftegaz, and TNK-BP. After the Rosneft purchase of Yuganskneftegaz, Moscow intended to merge Gazprom and Rosneft, but subsequently abandoned that effort. Rosneft's purchase of Yukos' assets at a rigged auction, however, clearly strengthens state authority and control of the energy sector.

Of the companies listed, most are notable for their absence or only minor presence in Caspian Sea fields. LUKoil is by far the most active and is committed to Caspian exploration and possible development. Gazprom and Rosneft are also represented, but not in Russian waters solely. The following partnerships are indicative—if perhaps not fully inclusive—of Russian companies' Caspian interests:

- In onshore Kazakhstan, LUKArco holds a 5 percent interest in the Tengiz field (LUKoil's interest is 2.7 percent), and LUKoil alone owns a 15 percent interest in Karachaganak, purchased from Gazprom.
- LUKoil and Yukos operated the joint venture, the Caspian Oil Company, for the possible development of prospects in the shallow offshore Russian waters of the Volga Delta.
- On the Russian–Kazakh median line, Rosneft and KazMunaiGaz agreed to develop jointly the Kurmangazy oil field with Total and Zarubezhneftegaz, the latter a venture of Gazprom and state-owned Zarubezhneft. Kazakhstan has jurisdiction over the field, but Russia has the right to develop half of it. Rosneft, together with Total, however, downgraded the field's profitability. Development was repeatedly postponed, a victim in 2004 of the more stringent Kazakh tax regime.[183] But in July 2005, Rosneft and KazMunaiGaz signed a PSA for the field's future development, even though it is a small field with estimated oil reserves of only about 135 million barrels.[184] Total may yet join the project, but has been reluctant to do so because of doubts about the field's profitability under the new Kazakh tax code.
- LUKoil and KazMunaiGaz are jointly to explore Khavalinskoye, which is primarily a gas field lying mostly in Russian waters. At present, it is believed to be a small field, with oil reserves of 264 million barrels, and condensate reserves of 125 million barrels. The field is under Russian jurisdiction, with KazMunaiGaz having the right to develop half of it.
- Tsentralnaya, also primarily a gas field, straddles the Russia–Kazakhstan median line. LUKoil and Gazprom formed the company TsentrKaspneftegaz to work with KazMunaiGaz on this field. The development of both these fields has been postponed for tax-related reasons, as well as gas market reasons. At one point LUKoil announced that it would start

production at Khavalinskoye in 2008, but that date has slipped to at least 2009–2010. The field's reserves are estimated at 521 million barrels of oil equivalent, so it is also a relatively small field. Gazprom has stated that Caspian shelf gas will be sold in Russia, but until the domestic Russian gas market is liberalized or gas from these fields can be exported, LUKoil is unwilling to proceed.[185]

- LUKoil is the sole developer in the Caspian Sea gas fields of Korchagina and Rakushechnoye, which lie fully in the Russian sector. Delay in development, however, is due to the same concerns as those that apply at Khavalinskoye.[186]
- In the offshore Yalama field that sits on the Russian–Azeri border of the Sea, LUKoil holds an 80 percent interest while SOCAR owns the remainder. LUKoil abandoned its only well in the field in 2005, but will continue development work.
- Elsewhere, LUKoil sold its 10 percent interest in the Azeri ACG fields, and declined to join the BTC oil pipeline consortium. LUKoil retains its 5 percent interest in the Shah Deniz gas field and the SCP gas pipeline through its venture with LUKAgip, which holds 10 percent in each of these. LUKoil terminated its 50 percent interest in the Zigh-Hovsan onshore/offshore field because of insufficient economic returns.
- LUKoil purchased Nelson Resources, a Canadian company, which owned the former ChevronTexaco interest in the Buzachi field.
- Finally, LUKoil has a 25 percent interest in Iran's Anaran block. The terms call for analyzing seismic data, conducting exploratory drilling, and if warranted, building transportation infrastructure.[187]

LUKoil's profile in the Caspian energy sector has grown rapidly in the last few years. As long as it continues to avoid clashes with the Kremlin, it will remain one of Russia's most successful and profitable private energy companies. Its Caspian interests, and those of Russia as a whole, however, are on the back burner of the overall Russian oilfield investment portfolio.

The Investment Climate

Russia's energy landscape has changed in recent years, as private companies like LUKoil and others have become major players and are run on commercial principles. As Moscow changes the old rules, and does not, however, fully define the new ones, uncertainties increase. The investment climate has dramatically cooled in recent years, largely due to the perceived politically motivated destruction of Yukos, once Russia's leading private oil company.

The energy investment climate was attractive to foreign investors during much of this time, most noticeably bringing in BP in a 50-50 deal with TNK, blessed by the Kremlin, in early 2003. But the investment climate rapidly soured later in 2003 and thereafter. The politics changed, Putin moved away from a democratic path, Russian taxes on oil companies rose, and the flagship energy company, Yukos, was fined and dismembered for fraud, tax evasion, and embezzlement in what is perceived as Kremlin retaliation against Yukos's jailed leader's political ambitions. According to a 2004 report by the Institute of International Finance's Equity Advisory Group, the government's seizure of Yukos's assets shook investor confidence "on a scale not seen since the Russian 1998 financial crisis."[188] Capital outflows in 2004 rose $33 billion, the highest level ever, according to the Fitch ratings service, much more than the $2–4 billion of earlier years, as a result of the more fragile and uncertain climate.[189]

Capital outflows do not tell the whole story, if capital inflows are ignored. Foreign interest in Russian oil and gas remains exceptionally strong, and while the Yukos affair caused a pause in interest, the promise of earnings from Russia's vast resources has since reversed the flow, and capital is returning. Foreign investment in the energy sector is at an all-time high and was up significantly in 2005 over 2004. In 2005 Gazprom secured a loan of more than $10 billion toward the purchase of a controlling interest in Sibneft. The loan was completely financed by a Western consortium of banks (the largest loan ever made to any Russian company in any industry). Rosneft likewise secured a $7.5 billion loan in 2005, again financed by a consortium of European and U.S. banks, including many of those making the loan to Gazprom. Banks are clearly returning to Russia, and major foreign oil companies with interests there are continuing to invest.

Foreign investors in Russia have realistic concerns, nonetheless, for the rule of law, property protection, and capricious taxation. The government currently takes 80 percent of oil company revenues (not profits) at prices of more than $27 per barrel. ConocoPhillips in 2004 acquired the Russian government's 7.69 percent share of LUKoil for nearly $2 billion, and has since raised its stake to 10 percent. That share could ultimately rise to about 20 percent, but probably not higher, given the Kremlin's desire to keep foreign companies in a noncontrolling position. In addition, Total of France owns a 40 percent share of an oil-producing venture with LUKoil. The Russian government—in part to try to offset the negative views of the international investment community after the assault on Yukos—explicitly approved both the ConocoPhillips and Total deals.

The foreign investment climate for Russian energy and other projects apparently deemed "strategic," however, has weakened. The government announced in 2005 that majority foreign-owned companies would not be permitted to bid for the rights to develop several large new fields, since these are now considered "strategic" assets. Firms permitted to bid must be 51 percent Russian-owned companies, although they may seek minority foreign interest. This means that TNK-BP is ineligible to bid unless the ownership restrictions are revised, or exceptions made; accordingly, the company is restructuring itself to meet the ownership ceiling. The new limit has dampened foreign interest and is another indication of how the Kremlin is seeking to extend its control over the energy sector.[190] In another move, Moscow also blocked in April 2005 the proposed Siemens purchase of OAO Power Machines, which makes both energy equipment and turbines for Russian nuclear submarines.[191]

Despite the authoritarianism and Putin's growing concentration of power, it is indisputable that Russia has made giant steps since the total collapse of the economy in 1998 and 1999, and the outside world has welcomed these economic and other improvements. In mid-2004 the IMF praised the growth and progress of the Russian economy.[192] The Fitch ratings service in 2005 upgraded Russian debt to BBB, which suggests that the economy-wide impacts of the Yukos affair have been minimal and short-lived. These external tributes to sound economic management should serve to reduce fears of Western investors, provided the Kremlin's political tightening and moves away from democracy do not offset them. Since his arrival first as prime minister and then as president, Putin has worked to lessen corruption (and charging and convicting Yukos's Khodorkovsky was part of that effort) and to build civil society, although the job remains unfinished. Russian foreign reserves are at an all-time high of more than $185 billion, the fiscal budget is in surplus, debt is being paid down, full liberalization of the currency is on schedule for 2006, and economic growth is expected to continue at approximately 5–7 percent for the rest of the decade. The country created a stabilization fund for oil revenues that reached $52 billion at end-2005.[193] Putin reduced income tax rates, and tax collections are up, although value-added taxation implementation is under fire as courts rule that companies cannot get VAT refunds on purchases from suppliers if they borrowed money to pay the tax.[194] Negotiations on Russia's entry into the WTO are proceeding. Military spending has declined.

Despite the overall economic progress, Russia is still viewed as a risky and uncertain business environment, and foreign companies are understandably cautious about committing huge amounts of capital. This cautiousness derives from their:

- Concerns over governance. The December 2003 parliamentary elections and the March 2004 reelection of President Putin underscore the movement away from democracy to a more authoritarian state with political power concentrated at the top. Rumors circulate that President Putin's backers will either seek to amend the constitution to permit his standing for a third presidential term in 2008, despite Putin's own statement that he will step down at the end of the current term. Companies may remain leery, nonetheless, of increasing capital commitments in a political environment where potential future leaders are not readily apparent.

- Concerns over the deteriorating situation in Chechnya, Ingushetia, and the North Caucasus in general, and the potential for violence throughout Russia. This instability poses a threat to the BTC and SCP pipelines to the south, and to the oil pipeline running from Baku to Novorossiysk.
- Concerns over protection of property and the rule of law. Press reports suggest that the court proceedings in the Yukos case demonstrate that the judiciary is subservient to the Kremlin.
- Concerns over the potential treatment of foreign ownership interests in Russian oil companies. It is President Putin who will decide, as he did in the ConocoPhillips and Total instances, how much of the Russian oil and gas industry foreign entities will be permitted to own. As Moscow moves to reassert itself as the dominant energy player, the TNK-BP venture has found it wise to invite Gazprom to join in the TNK-BP Kovytka project, underscoring the point that foreign interests cannot own a controlling share in Russian energy projects.
- Concerns over contracts already in place. The CPC project is a good example of this concern, particularly since the Russian government owns 24 percent. The project was designed for profitability at a maximum capacity of 1.3 mmb/d, with an initial capacity of 565,000 b/d. The pipeline has a life of 40 years, and it appears that despite a signed agreement to construct and operate it, negotiations will never end and will continue for the full 40 years. The pipeline's expansion plan has been an issue: Russia/Transneft did not want CPC to expand until it became profitable, while CPC held that only expanded capacity would bring profitability. Russia wanted CPC to raise tariff rates so that Moscow would benefit through higher transit fees, but CPC said rising rates would spur oil shippers to find other routes. This impasse long underlaid the attempts of Transneft to bring CPC under its direct control.[195] But in March 2005, all CPC shareholders finally agreed to expand the line to full capacity and raise rates from about $3.70 per barrel to more than $4.00.[196]
- Concerns over the inability to establish workable production sharing frameworks for developing oil and gas fields. Russia has long believed that it does not need PSAs to develop its resources, saying that the existing tax legislation is adequate to attract foreign investment. Companies typically prefer PSAs since such arrangements give them predictable tax regimes for the entire lives of projects. While PSAs in Russia have been dead since mid-2003, the country is now planning to revive them, but only for difficult new offshore developments, such as those in the Arctic, the Barents Sea, on Sakhalin, or those in the Caspian Sea that face geologic and technical hurdles.[197]
- Concern over the instability, arbitrariness, and unpredictability of the tax system, all of which create uncertainty and reduce business investment. The implementation of existing tax legislation is not well defined, with the government "determined to make sure that companies pay the

government's own interpretation of . . . taxes outstanding."[198] President Putin noted in 2004 that since 2000 the split in revenues was 54 percent for the government and 46 percent for the companies. He added that elsewhere in the world the split is closer to 80 percent for the government,[199] and he has taken steps to hike the government's tax take. Export duties are now based on a sliding scale and recalculated every 2 months, with the percentage increasing as prices rise.[200] Production taxes in 2004 rose 15 percent, and export duties on crude oil have nearly doubled from 2004 to 2005 to close to $18 per barrel.[201] However one parses the particular taxes, the bottom line is that Russia presently collects 90 percent of the value over a price of $27 per barrel on exported oil.[202] Despite these tax levels, the Russian public is thought to believe that both Western and Russian oil and gas companies are not paying their fair share, or else acquired assets on the cheap and should now pay back from the earnings.

- Concerns about the purpose of what appears to be Gazprom's capricious stalling of projects with international investors. The proposed $18-billion, 30-bcm/y Kovytka gas-condensate field development and pipeline project in Eastern Siberia is a case in point, and the major foreign investor is BP.[203] In a year fraught with uncertainty, TNK-BP was allowed in 2005 to keep its license for the Kovytka gas field, although the license issue could arise again in 2006.[204] The field contains estimated recoverable reserves of more than 2.1 trillion cubic meters, and would require a pipeline of more than 4,800 km to China. This line would join China's West-to-East gas pipeline. China National Petroleum Corp. and (South) Korea Gas Corp. signed letters of intent in late 2003 to buy 20 bcm/y and 10 bcm/y, respectively, for 30 years, with another 4 bcm/y for use in Russia. Production was intended to start in 2006, with deliveries to begin in 2008, but these dates will most certainly slip.

  But now Gazprom is considering incorporating Kovytka development as part of its plan for a unified gas supply system in Eastern Siberia to include Sakhalin gas development, which would mean a Kovytka pipeline not to China but through the Sakhalin region for export. China is pressuring Russia to approve the original Kovytka project, and has threatened to look elsewhere for future gas supplies, including LNG, if approval is delayed or denied.[205] The longer Russia stalls on gas sales to China, the more leeway and encouragement Kazakhstan sees, as it also seeks to sell gas to China. TNK-BP has likewise said it will withdraw its support should the project fail to win approval. The stalemate could yet end the project, as China and South Korea tire of waiting and take their business elsewhere.

- Concerns over past or good-faith agreements. The Russian government in January of 2004 cancelled the 1993 tender at Sakhalin-3, granted by the previous government, that gave ExxonMobil and ChevronTexaco the rights to develop the acreage, together with Rosneft, under PSA terms.

> The formal license was to be issued after the signing of the PSA, but the signing never occurred. Sakhalin-3 lost its PSA status in 2003 with the change in the Russian tax code. While ExxonMobil said at the time that it could accept non-PSA terms, the company is likely to lose the project if the acreage is retendered under the Putin government, as is expected in 2006, with China's Sinopec likely to win a share of the tender.[206] Sakhalin apparently falls in the category of fields considered "strategic," where foreign firms can own no more than 49%, as announced in 2005. A new licensee(s) would presumably be required to compensate the company for its $65 million in exploration expenditures.[207] In 2004, the American Chamber of Commerce in Moscow viewed the cancellation and the invalidation of the Exxon license as tantamount to expropriation.[208]

Despite these concerns, high oil prices fuel investor interest, both Russian and international. Production is expanding, as are transport systems and port development. The pipeline system and port of Primorsk, for example, is being expanded to provide export capacity of 1.2–1.3 mmb/d. Until the arrest of Yukos's Khodorkovsky, who promoted the Murmansk pipeline and port complex, that project looked promising. With Khodorkovsky's arrest, Russian interest ebbed, as has Russian interest in the oil pipeline from Western Siberia to China that Yukos also proposed.

For large and long-run projects, nonetheless, Moscow knows that its own firms may have neither sufficient capital nor sufficient technical depth. The government promoted discussions between Gazprom and ConocoPhillips to join forces in a 20-bcm/y LNG project at Murmansk, with half the output slated for U.S. markets and the other half for European ones. This is a separate project from the Yukos Murmansk oil pipeline and port project, which called for oil export capacity via large tankers to reach 1.6 mmb/d, destined largely for the United States.[209] Neither of these Murmansk projects, were they to come to fruition, are near-term ones. Oil ports and pipelines could probably, however, be realized reasonably soon, in perhaps 5–10 years, if Transneft supports them. The LNG proposal to use gas from the large Barents Sea Shtokman gas field, however, is not expected to happen for years; 2020 might be a relevant target year. Knowing that Gazprom has no experience with LNG construction, Putin has already met with British, Norwegian, and American companies with LNG experience. Gazprom subsequently announced that its short list of possible partners to develop the Shtokman field include Statoil, Norsk Hydro, Total, ChevronTexaco, and ConocoPhillips (no inclusion of ExxonMobil or Shell, perhaps because of difficulties at Sakhalin). Future Russian LNG sales to the United States will in part depend on the anticipated size of U.S. LNG markets as well as the locations of new U.S. receiving terminals.

Russia and OPEC

Russia's production of Caspian oil and gas is likely to remain a small share of its total production, even if a number of the jointly managed fields go forward in the years ahead. Any Russian Caspian oil production, however, may suffer if OPEC in the future drives down oil prices.

In a period of relatively high oil prices, we expect Russia to ignore OPEC, and to produce at maximum capacity. Similarly, if prices plunge as they have done in the past, we likewise believe Russia will ignore OPEC calls for output cuts. This will be as much a political decision as an economic one. Having taken the reins of the energy industry, Putin is not about to cede decisions on Russian oil and gas to OPEC, many of whose members would be delighted to cheat on production quotas and expand output at Russia's expense. In fact, a growing and modernizing Russian energy industry could diminish OPEC's role in world oil markets, while contributing at the same time to promoting the use of natural gas.

Russia and its companies are price takers and revenue maximizers and believe that production of 9.5 mmb/d in 2005, and a target of 12 mmb/d for the future, will continue to command top prices, subject to quality differences, in a world market of more than 83 mmb/d. Yet Russia's increasing output in itself may bring about an oil price collapse, as production from it as well as from Canada, North and West Africa, South America, and the Middle East mounts over the next decade and beyond. In a period of depressed prices, as was the case in 1999, OPEC approached Russia and asked that it cut production to support prices. Russia ignored the request, but at the time it was a less significant player, producing at a low of about 6.2 mmb/d, of which less than 4 mmb/d of crude and products were exported. Russia has since recovered, primarily because of the devaluation of the ruble in 1998 and higher oil prices since 1999.

OPEC's key member, Saudi Arabia, once again approached Russia in the fall of 2003, looking for cooperation on output levels and oil prices. While President Putin and then–Crown Prince (now King) Abdullah signed a 5-year agreement of cooperation, it seems not to have amounted to much.[210] Russia correctly maintains that it cannot control production of its private companies the way OPEC does of its national companies. OPEC remains sceptical of the truth of this statement, recognizing that with Putin's consolidation of central governing authority since the 2003 and 2004 elections, the Russian oil companies are under his thumb. This means that the companies would be loath to ignore a presidential request to reduce production, should Putin deem this necessary to avoid a full-fledged price war. Nonetheless, despite the pain that would be inflicted from falling prices on Russian revenues, Russia correctly believes it is better positioned, with its more diversified economy, to ride out a price war than are the Saudis and other OPEC members. As long as prices remain high, Russian companies will produce to the limit to take full advantage of the income flow. This strategy of course benefits the government as well, through the tax take.

Russian oil companies including both Rosneft and LUKoil responded as expected to OPEC's decision to cut oil production by 1 mmb/d beginning April 1, 2004. Neither had any plans to reduce production; rather, they both planned to expand it. A Russian news agency quoted anonymous energy ministry sources that said OPEC's cuts allowed more room for Russian oil in the market.[211] Caspian producers also welcomed the OPEC cuts, for Caspian producers were also the beneficiaries of additional room in the market and relatively higher prices than otherwise.

The forecasting of prices and of quantities supplied and demanded is a difficult and uncertain business. With prices hitting more than $60 per barrel in mid-2005, some forecasters see prices continuing to rise in 2006 and, thereafter, pushed upward by strong world demand. Other forecasters are projecting that both Brent and WTI per-barrel oil prices will plummet to lows of $20–30 in the near future, as a result of new supplies to the market overwhelming any pickup in demand. At that price level, Russian companies would largely curtail or abandon shipments by rail and barge, for these are widely believed to be unprofitable at $20. Low price projections, if borne out, could be damaging to Russia, since oil earnings are the main source of funds for the budget. While Russia's cooperation with OPEC, in the face of a severe drop in price, and a similar severe drop in revenues to the state, cannot be ruled out, we think the probabilities are low since cooperation with OPEC would impugn Russia's desire to be viewed as a reliable supplier.[212] Our view is that Russian companies will produce full out whether prices are at $60 per barrel or at $10, provided costs are recovered. The companies as revenue maximizers will ride the crest of expanded demand at the lower price levels, and not be cowed by the demands of OPEC.

About the only thing the Saudis could do to try to teach Russia a lesson would be to expand production and drive the price down below Russian variable costs. What happens to Russia's Caspian production, as well as other Caspian production, in the event of an all-out price war between OPEC and Russia? Clearly Russia will be hurt, but undoubtedly far less than OPEC countries, whose economies are far less diversified than Russia's. Certainly Russian Caspian production would be severely curtailed and could even close down, depending on how long the price war were to last. While cost information is not widely available, we know from company representatives that at a price of about $9 per barrel, no further Caspian investments will occur. At prices below that level variable costs may not be covered. Caspian oil—with its depth under the seabed, the complexity of the geology, and the presence of impurities such as sulfur and mercaptans—is high cost production. Much as the companies would resist walking away from billions of dollars in investment, they will not produce at below-cost price levels.

Other Issues

There are several additional considerations that influence the development of Russia's Caspian oil and gas interests. These include swaps and pipeline transportation, the monopoly status of the Russian gas and oil transport companies, and the political engagement with the U.S. government and private sector "Energy Dialogue." We take a brief look at each in turn.

First, Iran continues to seek Russian oil to swap at Persian Gulf ports and to push for an oil pipeline south through its territory. Although Tehran may face some loss of future swaps with the opening of both the BTC line and the Kazakh line to China, it hopes nonetheless to continue to attract Russian oil for swap, be it produced in the Caspian or elsewhere. Russia has built up its pipeline and export terminal capacity in the north, permitting some shipments to move by pipeline rather than be swapped, and thus leaving Iran potentially with a large system to receive swaps but no takers. If swaps fail to endure, could a pipeline to an ocean-going port attract joint Russian–Iranian interest? Iran maintains that it offers the safest and cheapest route for Caspian and Russian oil to reach Persian Gulf ports. Despite U.S. policy toward Iran, it is possible that Russia and Iran could in the future partner in building this route, although issues could arise in defining the role of Transneft. A trans-Iran oil pipeline is far less costly than either the Murmansk pipeline project, now moribund, or the proposed $12 billion oil pipeline from Western Siberia to the Russian coastal port of Nakhodka.[213] Politically, the United States has little ammunition to prevent such a project without risking serious deterioration in its relations with Russia.

Second, we expect the Gazprom and Transneft Russian gas and oil state-owned transport monopolies to remain monopolies for some time to come, although their monopoly status and roles could become a major issue in Russia's bid for membership in the WTO.[214] Given the sea change in Russian politics, already resulting in a stronger role for the state in the energy sector, we do not expect any reform efforts directed at permitting competition with either Gazprom or Transneft to occur under Putin. Gazprom has so far countered attempts by some Russian companies to break its monopoly position, and has increased its efforts to gain and maintain control over export pipelines in Belarus and Ukraine. It has also insisted on retaining all gas production from its one and only joint production venture with the German company, Wintershall. It has as well joined forces with Russia's state-owned Rosneft to develop oil fields in East Siberia for export to China, which some observers believe is a tactic to reduce Western firms' interest in Russia's east.

In fact, the decision has been made that there will be no private pipelines in Russia, except for two: in addition to CPC as an exception to the "no private pipelines" rule, the second exception will be the oil pipeline serving the Sakhalin fields in the East. All others will be built either by Gazprom for gas or by Transneft for oil.[215] Both will remain monopoly and majority state-owned companies, although foreign shareholders are expected to be limited to holding no more than 20 percent of the shares.

Gazprom is also unlikely to yield its gas monopoly. We believe it will retain firm control of all Caspian gas sales to Russia and of the transit revenues, unless and until new gas pipelines are built beyond Russia's control. Decisions of this nature are not made quickly, as the history of both BTC and CPC demonstrates. While the SCP gas line from Shah Deniz in Azeri waters to Turkey is under construction, gas producers to the east of the Caspian Sea witnessed the failure of the trans-Caspian gas project because of the inability of Turkmenistan and Azerbaijan to agree on terms. Uzbekistan, Kazakhstan, and Turkmenistan are likely to wait years to sell gas to China or to have alternative outlets through Iran. In the meantime, Gazprom retains its monopoly position.

Transneft likewise will retain its monopoly status over oil transport. It has reacted heatedly to the mere suggestion of any private pipelines in Russia. In that regard, it is trying to change the rules governing CPC so that Transneft can control CPC tariffs, shipping, and capacity decisions. Transneft has opposed the ideas of private pipelines to China, Japan, and Murmansk. Yukos proposed both the line to China and the one to Murmansk. Yukos was joined by TNK-BP, Sibneft, and Surgutneftegaz for the Murmansk line, estimated to cost $6–7 billion and designed to carry 1.6 mmb/d.[216] With the Murmansk complex aimed in part at serving the U.S. market, Transneft was allegedly seeking U.S. government purchasing guarantees, perhaps as a means of retarding private company interest. With Yukos now out of the picture, both routes appear shelved.

As with gas, any oil that leaves the Caspian via Russia as an intermediate country is subject to the terms imposed by a monopsonist. The key exception is the private CPC, originating in Kazakhstan and ending in Russia on the Black Sea. If Putin withstands Transneft's desire to change the rules on CPC, then the seed of pipeline competition will be planted as Transneft recognizes that shippers have a choice. With the commencement of BTC deliveries in 2006, a significant dose of competition is likely to follow, in that oil that used to flow through Russia now has an alternative route. This may yet lead to Kazakh shipments on BTC, especially after 2010 or 2012 when Azeri volumes are likely to fall sharply.

In a new development to involve Transneft, Georgia's president Saakashvili proposed the construction of a pipeline southward from Novorossiysk through Georgia's restive province of Abkhazia to link with the BTC pipeline. He expects that with the opening of BTC, the Turks will take action either to severely limit or to close the Bosporus to oil tanker traffic, despite the 1936 Treaty of Montreux.[217] This would leave about

1 mmb/d of Russian, Azeri, and Kazakhstan oil stranded at Novorossiysk with no Black Sea outlet other than Odessa, which will not serve as an outlet as long as shipments flow south from Brody. Such a decision would certainly underscore the urgency of a Bosporus bypass. If the suggested line to Georgia were built, tariffs would be higher to Ceyhan than to Novorossiysk. As long as political difficulties in Abkhazia continue, most observers give this pipeline proposal little chance of realization. We view the likelihood of Saakashvili's idea being executed as extremely low.

For the Caspian producing region, the continued monopolies of Transneft and Gazprom on both existing and possible new pipelines within Russia are a grave concern. The Caspian producers can solicit all the international partner help and financing possible, but if any such lines actually come into existence, and must tie into the Russian system to move to international markets, Transneft or Gazprom will remain all-powerful in making price and capacity decisions.

Third, despite the United States and the West's concern for Putin's choice of direction for Russia, and recent energy developments there, Moscow and Washington have continued their Energy Dialogue begun in 2002.[218] That process established both government-to-government groups, as well as Russian–American private company groups. The government and company meetings have occurred on parallel tracks to develop recommendations to improve the investment climate. The goals of the Energy Dialogue contained in the original joint statement by Presidents Putin and Bush were to

- "Develop bilateral cooperation in the energy sphere on a mutually beneficial basis in accordance with our respective national energy policies.
- Reduce volatility and enhance predictability of global energy markets and reliability of global energy supply.
- Facilitate commercial cooperation in the energy sector enhancing interaction between our companies in exploration, production, refining, transportation and marketing of energy, as well as in implementation of joint projects including those in third countries.
- Encourage investment aimed at the further development and modernization of the fuel and energy sector of Russia, including expansion of oil and gas production in Eastern Siberia, the Far East, and offshore areas.
- Promote access to world markets for Russian energy, including through the commercial development and modernization of Russia's port and transportation infrastructures, the electric power and gas sectors, and oil refining capabilities.
- Foster science, technological, and business cooperation in the use of unconventional energy sources, and energy-efficient and environmentally clean technologies.
- Cooperate in elaboration and development of new ecologically safer nuclear power technologies."[219]

One of the key areas for discussion has been the identification of barriers to investment in the Russian energy sector, but developments with the Kovytka gas project, Sakhalin-3, and Yukos have slowed progress in the last few years. The communiqué issued after the Bratislava summit in February 2005 suggested, however, that there are some specific projects to provide the basis for further talks.[220] It is clear that with oil prices at more than $60 per barrel, however, Russia feels no urgency to engage in oil cooperation with the United States, but it needs external assistance with proposed LNG developments.

The visit of U.S. Energy Secretary Samuel Bodman to Moscow a few months later in May 2005 demonstrated U.S. interest in continuing the Dialogue, and representatives of both countries met the next month. Since then, Russia publicly expressed its interest in cooperation with the United States on future LNG development and LNG trade, and Putin met with the heads of at least three U.S. majors who are interested in investing in the Shtokman gas field in the Bering Sea, a likely source of LNG exports to the United States. That ChevronTexaco and ConocoPhillips are on the short list for potential Gazprom LNG partners is a great boost to the Dialogue's progress. The United States is also interested in encouraging a northern oil export pipeline to Barents Sea ports, where oil could be exported year-round to U.S. and Western markets.

Although the U.S and Russia have since 2002 signed a protocol on energy efficiency and renewable energy, and another on oil spill prevention and response, few additional protocols have been agreed. The energy working group (EWG) includes subgroups on the oil market, on investment, on energy efficiency technologies, on information exchange, and on small and medium enterprises. Decisions to cooperate or not will be made only with express approval by the Kremlin, regardless of the existence of joint U.S.–Russian working groups or commercial energy summits. The governments on both sides can provide the environment for cooperation, if they truly wish it, but unless individual companies on both sides also take the initiative, the Energy Dialogue will die.

Conclusion

Russia's oil and gas reserves, because of their size, will remain attractive to foreign investors, provided projects are economic and the fiscal and legal terms are known and not changed at will. With the reelection of President Putin and his tightening of the reins on the energy industry, the investment climate deteriorated, but high prices and ongoing foreign interest in Russia's supplies are leading international investors to put these setbacks behind them and to return. More major projects such as the BP 50-50 joint venture with TNK are not on the horizon, but minority holdings are, as appears to be the case with the proposed development of the Shtokman gas field. The terms are expected to be similar for projects in the Russian Caspian or in the shared Caspian border fields.

THE INNER CIRCLE: CONCLUSION

The prospects and outlook for investment in and development of energy resources in the Caspian region looks brightest for Azerbaijan and Kazakhstan, although uncertainties in the investment climate especially plague the latter. Not only do both countries have reserves of oil and gas, but they also have shown interest in working with foreign investors to exploit those resources, despite the inevitable conflicts and disagreements. The major and tightrope issue that both the governments and the participating companies face is how to divide the rents. If the governments demand too much, which may sell politically at home, the investors will find the investment climate unfriendly and may decide to reallocate their capital elsewhere. If the foreign companies, rightly or wrongly, are perceived as exploiting the resource for their own sole benefit, they are likely to be subject to proceedings, which, in turn, could deter their interest in continuing operations. Neither side wants either of the extreme alternative outcomes. As a result, despite the presence of signed contracts in place, negotiations are likely to be an ongoing process, with each side trying to tilt or rein in the other in the interests of meeting its objectives.

A detailed look at the prospective and producing fields substantiates our conclusions on forecast oil output, particularly in Kazakhstan and Azerbaijan. We do not expect the former to reach 3.5 mmb/d of oil production by 2015 as the leadership has so often proclaimed. Nor do we expect Azerbaijan to be able to maintain its future production level of 1.2 mmb/d for any sustained length of time. In brief, both potential producers and potential consumers have heretofore exaggerated the production profile for reasons of wishful thinking and political gain.

The notion of equitable division of the rents is one that has not registered yet in Turkmenistan, which lags behind the former two countries in exploiting its energy reserves, which are primarily natural gas. Moreover, Turkmenistan's energy future suffers from a dearth of transport alternatives to the former Soviet system. This means it remains subject to monopsony power in the hands of Gazprom.

Russia and Iran have not pursued development of their Caspian region oil and gas reserves with the speed or intensity of the other littoral states. While Iran has to date been largely bypassed in the development of Caspian energy resources owing to its political upheavals, disputes with the West on nuclear matters, and intransigence on Sea delimitation, it has every intention of reasserting its historical role in management of the Sea and its oil and gas. Iran's inaction is also due to the fact that it has vast oil and gas reserves elsewhere in the country that are easier and less costly to develop.

Nonetheless, in a show of its interest in the Caspian, it is pursuing oil-swap contracts at its northern coastal ports with Russian and other regional producers for equivalent barrels delivered to Iranian Persian Gulf ports, and it has announced that it will go ahead with exploration of its uncontested portion of the Sea.

In Russia's case, Gazprom and LUKoil are well aware of the possibilities of Caspian production, and are positioning themselves for future planned development activity, even if their known reserves in the Caspian region remain speculative at present. Russia will continue to exert its monopsony power and pay less than international prices for Caspian gas until the Caspian countries arrange alternative gas export routes, for it is in Russia's interest to keep Caspian producers dependent on the existing Russian pipeline network. At the same time, Russia views Caspian oil and gas producers as future potential competitors for its own sales of both oil and gas to world markets.

4

The Outer Circle

A number of countries outside the inner circle are interested in energy and political developments in the Caspian region. We limit our consideration to China, Uzbekistan, and the United States, and in this chapter address each in turn. Other countries with expected rapid growth in energy demand, such as India and Pakistan, may also express interest in energy developments in this region, but for the foreseeable future they have adequate oil and gas supplies, both internally and from other exporters.

China is increasing its interest in the current and future production possibilities of this region, and is looking to Central Asia, Russia, and other areas for diversified future energy supplies. Its energy demand has grown spectacularly, and is likely to continue to rise, but the growth rate will depend on whether China's unsustainable economic growth results in a "hard landing." If that occurs, GDP growth will slow or fall, resulting in reduced energy demand in the near term. Beijing is nonetheless scouring the world for new oil supplies, looking for alternative sources beyond the Middle East, for it fears overdependence on what it considers an unstable area. Given the size of Russia's hydrocarbon resources and China's economic and energy growth, energy trade between the two countries should be promising, but contracts have been elusive. Uzbekistan, while rich in gas, is essentially a nonplayer in international energy markets, given its Neanderthal politics and unwillingness to open its economy or permit any kind of nascent democratic expression. For its part, the United States has publicly and repeatedly favored pipeline projects that avoid both Russia and Iran as part of its "multiple pipelines" policy, yet recently advocated the proposed oil pipeline project within Russia to Murmansk, which would permit exports from a year-round deepwater port. U.S. interests in the Caspian area now appear, however, to be driven more by the war on terrorism than by either its interests in energy supply diversification or in building democracy.

CHINA

While the energy sector is driving economic growth in Russia, that sector could, in the future, be the growth-constraining factor in China. China's energy demand has risen rapidly in the last decade, as the country moved from a position of self-sufficiency to that of a net crude oil importer in 1993. As a result of its recognition that crude oil import diversification is a sound policy, the country is looking worldwide for additional sources of oil, both to buy and to develop.[1] To the extent that Central Asian oil and gas, as well as Russian supplies, could also move east on either new or existing pipelines, energy exports to China permit both the importer and the exporters to realize gains from trade.

Bordering Central Asia's eastern frontiers, China is extremely interested in the prospects for Caspian oil and gas development and their possible export to China. The country's announced goal is to quadruple its GDP by 2020, and whatever success it achieves in reaching this goal will necessarily involve greater energy use, including both oil and gas. China is the world's fastest-growing oil importer; its imports grew by nearly a third in 2004, and amounted to 3.2 mmb/d in 2005. Some experts, however, viewed China's oil demand growth in 2004 as unsustainably high at 978,000 b/d, and noted that growth was stagnant in 2005, as the building of oil inventories declined and more coal-fired thermal plants came on line to meet new electricity demand.[2] China's strong economic and energy growth, together with the events of 9/11/2001, have reconfirmed to Beijing that the Middle East is a region of never-ending instability, and one on which it would be unwise for China to increase its dependence.[3]

Concern for Energy Security

China is looking for ways to increase its oil supply security to permit its economic growth to continue. Given the robust forecasts for Chinese energy demand growth in the next decade and beyond, the country's leadership adopted in 2003 an energy strategy more than 10 years in the making. The strategy includes increasing productivity at existing producing oil fields in China, establishing a strategic oil reserve, encouraging conservation and renewable energies, increasing the use of natural gas, looking for ways to diversify oil and gas future imports away from the Middle East, and securing more producing prospects abroad, including in the Caspian region. As a result of its rapid growth in oil demand, China imports more than 50 percent of its oil from the Middle East, despite its efforts to secure equity production interests elsewhere.[4] Nonetheless, the equity interests of Chinese energy companies have so far contributed little to the country's energy security, providing only minimal supplies, which would do little to offset any supply-disruption losses in the Middle East. While the strategy calls for diversification away from the Middle East, the reality is that Middle East imports will continue to loom large in China's future energy oil and gas imports.

The key components of China's energy strategy include the construction of a strategic petroleum reserve, the encouragement of a variety of new pipelines, and the acquisiton of equity interests in oil and gas producing assets overseas. The latter goal was most forcefully demonstrated twice in 2005 by CNOOC's surprise and unsuccessful bid to buy Unocal for $18.5 billion, and by CNPC's successful bid to buy PetroKazakhstan for nearly $4.2 billion. Both Unocal and PetroKazakhstan, however, were small players in the Caspian, with the former owning only about 10 percent in the ACG fields (or appoximately 120,000 b/d of proposed peak production) along with 9 percent of BTC pipeline capacity, and the latter producing only about 150,000 b/d. Beijing, nonetheless, is pursuing its energy strategy on all fronts:

- China is building a crude oil strategic reserve of 100–350 million barrels. The first of four storage sites (possibly increasing to six sites), with a capacity of 31 million barrels, was completed in 2005. Filling is to begin in 2006, building to a capacity of 100 million barrels. The other three sites are expected to be completed in 2007 and 2008. Depending on the rate of fill, the reserve may cover 35 days' worth of imports in the first quarter of 2007, as much as 50 days of imports by 2010, and 90 days of import cover by 2020.[5] What is uncertain is whether China will move ahead with robust fill rates at current relatively high crude oil prices, or delay its fill schedule in expectation of lower future oil prices.
- Some three-fourths of China's oil imports currently pass through the Malacca Straits, frequently the target of pirates and terrorists. Should ships carrying Middle East oil suffer accidents or sinkings, the country's imports could be severely disrupted. To offset this risk, China supports not only an oil pipeline from the east from Kazakhstan, but also other routes. One of these is a pipeline originating at Sittwe on the west coast of Myanmar, fed by tankers carrying Middle East oil, with its terminus in China's Yunnan province adjacent to Myanmar. This route is shorter and safer than moving oil via the Malacca Straits, but more costly per mile than the line from Kazakhstan.[6]
- Other major oil pipeline ideas under consideration that avoid the Malacca Straits are a line from Karachi, Pakistan, to China's western province of Xinjiang, and one from Chittagong, Bangladesh, through India and

Tibet. Both would be fed by seaborne delivervies of Middle East oil at their points of origination. While China and Myanmar have no history of political disagreements, that is not the case between China and India. Both of these pipeline ideas have not been fully costed out. Both are likely to be expensive, however, because of weather conditions in the first case and high construction costs in the second.

- In the Caspian region, Chinese companies are particularly active in Kazakhstan and Azerbaijan. In 2005 China signed an oil and gas development agreement with Uzbekistan.

The data suggest that the country's concern for future energy supplies and security is well placed. In 2005 the country's production of about 3.6 mmb/d was insufficient to meet the demand of 6.9 mmb/d. Domestic output is either flat or declining, and is expected to drop to 2.2 mmb/d by 2015, although the government is actively pursuing steps to reverse this slide. With economic growth of about 9–10 percent per year, oil demand grew at 15 percent annually in recent years, although there has been some slowdown as the authorities have sought to constrain economic growth.[7] With 2005 crude oil imports of about 3.3 mmb/d expected to increase (absent a downturn in economic growth), China's crude oil imports could exceed 5 mmb/d by 2010, and 10 mmb/d by 2020. Some China watchers forecast that by 2030 China's oil import demand could rival that of the United States.[8] Estimates vary as to whether demand will rise at roughly 3.5 percent annually to 2030, or at 6–7 percent per year. The key determinants will be how rapidly the overall economy grows, whether economic growth can be sustained or it experiences sharp downturns and contractions, and how these growth rates affect energy demand.

Prospects for Economic Growth

Infighting among Chinese politicians and economic experts on appropriate macroeconomic policy continues. In the short to medium term, the Chinese economy and its growing energy demand are at risk of a "hard landing" as Chinese officials try to dampen the current and unsustainable economic growth rate. While there are reports that the country's economy is slowing, many observers estimate the current growth rate is more than 9 percent.[9] (Some think that the true growth rate in recent years was closer to 15 percent.[10]) While others believe that a hard landing is inevitable, some think China may yet be able to avoid it.[11] A hard landing is generally taken to mean a steep drop in economic growth leading to higher urban unemployment and a sharp falloff in imports. It could mean overcontraction of the economy to below the 7 percent rate needed to provide jobs, while a soft landing would be achieved with a slowdown in growth to 7 or 8 percent annually. China has a record of cyclical growth characterized by rapid expansion, and then contraction, and energy demand rises and falls accordingly. But in the longer term, China is for the most part expected to continue its economic growth, if not at such a torrid pace as in recent years. This inevitably means it will be forced to make choices for the longer term and, as part of its energy strategy, to continue to diversify its oil and gas imports.

The forecasts for Chinese economic growth vary, in part because of the perceived unreliability of China's economic statistics. The Economist Intelligence Unit, for example, puts the likelihood of a hard landing at 40 percent, while Deutsche Bank experts are uncertain as to whether China will experience a hard, soft, or no landing at all.[12] China's decision to establish a crawling currency peg and to let the 11-year-old pegged rate of 8.28 yuan to the dollar rise 3.1 percent in July 2005, while raising the prices of its exports, will not do much in the short run to cool the country's growth or affect the probabilities of a hard or soft landing, unless China subsequently adopts a floating exchange rate. Moreover, China's statistics are often unreliable, with the authorities tending to overestimate the data for the good times and underestimate it in downturns. Morris Goldstein and Nicholas Lardy at the Institute for International Economics suggest that the landing will be long in any event, and growth may decline as much as 4 to 5 percentage points from its current rate.[13] In contrast, Peter Zeihan (writing in *Stratfor* in March 2005) suggested that China's economic growth miracle is about to implode, as the problems of the banking sector and state-owned enterprises mount and appropriate steps to address these issues are avoided.[14] Another author suggested in July 2005 that perhaps as soon as the end of 2006, the Chinese economy will hit a wall, the government will walk away from its WTO obligations, and foreign investors will flee, thus bringing about a hard landing and falling oil demand.[15] The government appears to recognize that excessive outstanding bank credit and real investment result in excess capacity in manufacturing and construction, worsen the bad debt problems of both industries and banks, and could lead to a sharp fall in growth and a concomitant rise in unemployment. While the latter is a potential contributor to civil unrest, the government has yet to take sufficient steps to prevent this outcome. The authorities, however, do appear to recognize the need to curb the unsustainable annual growth of 10 percent. The investment bubble, fed by what some call "irrational exuberance," may nonetheless be too large for normal economic measures to rein in and at the same time achieve a soft landing. Outside observers typically doubt both whether the Chinese have the macroeconomic tools to slow the economy to a sustainable growth rate without overshooting, and also whether they will show the determination to use the tools that are available.

Should China experience a hard landing, its demand for both oil and gas is likely to drop abruptly, if only for the next several years. Oil and gas demand in China's Asian trading partners—who owe their revived exports to China's growth—is also likely to contract if China has a hard landing. Any drop in economic growth will cause oil demand to decline, thus reducing upward pressure on oil prices. About the most accurate statement that can be made about China's oil demand, depending on the landing scenarios, is that the run-up in its oil imports in the last few years could be just as dramatically reversed. This means that some 1.8–2.0 mmb/d could be cut from world oil demand owing to a hard landing. Those who model China's economic growth and energy demand have concluded that for every percentage point decrease in the country's growth rate, the world oil price declines from $1 to $2 per barrel, provided OPEC does not take offsetting steps to tighten supply.[16] Accordingly, a drop in Chinese economic growth from 10 to 7 percent per year could take as much as $6 off the world per-barrel price of crude oil.

In its attempt to find non–Middle East sources of oil and gas to support its economic growth, China is scouring the world, even to entering bids for Canadian syncrudes that would normally be exported to the United States. To win increased Russian supplies, China has called for a tripling of bilateral trade, has supported Russia's candidacy for WTO membership, and has been flexible on its periodic border disputes with Russia.[17] Russia nonetheless severely disappointed it on two counts: the vetoing of the Yukos oil pipeline proposal to deliver East Siberian crude oil to Daqing and the delaying of the Kovytka gas development project.

At the end of 2004, Russia announced it would go ahead with a 4,100-km East Siberian oil pipeline, to be partially financed by the Japanese. This line will lie fully in Russian territory and terminate at the Russian port of Nakhodka, allowing exports to many buyers. Initially, China thought it had a deal with the Russians to include a 600,000-b/d branch line to Daqing, and was bitterly disappointed when the $12-billion Nakhodka project did not include a spur to Daqing. Russia presumably wanted full control over any pipeline to its east, and did not want to tie itself to China as its sole buyer, a lesson it learned from its difficulties in selling gas to Turkey on the Blue Stream pipeline. The story, however, is far from over, as a Russian adviser to Putin announced in April 2005 that Russia would, after all, build a spur line from Skovorodino to Daqing, but the adviser did not commit to the construction or to a construction date.[18]

Prospects for the route to Nakhodka are still uncertain.[19] A variant calls for building the line in two stages. The first stage, estimated to cost $6.5 billion, would include an eastward 2,000-km leg from the Angarsk–Taishet area through Russia to Skovorodino (70 km north of the Chinese border). Targeted for completion by late 2008, this line would be followed by the possibility of building a spur line from Skovorodino directly south to Daqing. This configuration offers China the same capacity as the former Yukos proposal, had the company received permission to build the Angarsk–Daqing line. The second stage from Skovorodino to Nakhodka as yet has no construction timetable, and its capacity has been lowered to 1 mmb/d from 1.6 mmb/d to reduce costs. Japan allegedly felt snubbed for its offer of up to $7 billion in financial assistance and disappointed by the uncertainties surrounding the line to Nakhodka.[20] Japan informed Russia that if Transneft builds the line to Daqing before the extension to the Pacific, it will withdraw its offer of financial assistance. Russia remains unmoved, stating that it does not need Japanese loans in view of other willing lenders.[21]

At this point it remains unclear whether and when China may succeed in tapping either Russian oil or (Russia's Kovytka) gas by pipeline. In the meantime, not wishing to wait on changing Russian politics and routes, China is stepping up its discussions with Central Asian countries for future oil and gas deliveries.

Chinese Companies in the Caspian Region

As China has moved to expand its sources of oil imports, its energy companies have become increasingly active around the world, and are playing a growing role in the Caspian region. Despite some private shareholders, these companies are all majority state-owned. They include CNPC (and its three subsidiaries PetroChina, CNODC, and CNPC [Hong Kong] Ltd.) as well as CNOOC, Sinopec, and Sinochem. These companies hold minority stakes at known Caspian fields, thus demonstrating their aversion to risk and their unwillingness to commit capital to new and untested prospects. Chinese companies failed in their one attempt to date to secure an interest in the new large offshore, and technically challenging, Kashagan field in Kazakhstan, but won the bidding war to buy PetroKazakhstan. While Chinese production is growing in the Caspian area, the volumes are by no means sufficient to permit China to reduce its imports from the Middle East. Until the opening of the pipeline from Atasu in Kazakhstan to China, production was either sold to Russia, refined locally, or sent by rail to China. Without a gas pipeline serving China from either Central Asia or from Russia, the country cannot realize its goal of increasing its use of natural gas as it attempts to diversify away from oil and coal.

Chinese companies are particularly active in Azerbaijan and Kazakhstan, but recently signed an agreement with Uzbekistan:

- In Azerbaijan, CNPC and its subsidiary CNODC participate with SOCAR in two small onshore fields undergoing rehabilitation (the Kursangi–Garabagli and Gobustan fields). ShengLi (a subsidiary of Sinopec) works with SOCAR at the onshore Pirsagat field. ShengLi, in partnership with Dubai-based Middle East Petrol, formed Noble Sky, and also joined with SOCAR in a PSA to rehabilitate the onshore Garachukhur field. Future production from these projects is unlikely to make a major contribution to meeting China's oil import needs; target production from all is expected to be less than 40,000 b/d. Output will be exported either through swaps with Iran or by pipeline to the Black Sea or to Ceyhan.

- In western Kazakhstan, CNPC acquired both ChevronTexaco's and Nimir's interests in the onshore North Buzachi field, giving it full ownership of that field. It subsequently sold half to Nelson Resources of Canada. North Buzachi is expected to produce 8,000 b/d, to be divided equally between CNPC and Nelson Resources,[22] the latter becoming a subsidiary of LUKoil in 2005. CNPC also acquired Nimir's 100-percent ownership share of the onshore North West Buzachi field. These fields contain heavy oil and were of marginal interest to western companies. CNPC and Aurado of Canada had signed a memorandum of understanding for CNPC to farm in on the Liman onshore block north of Atyrau, but the Kazakh government cancelled Aurado's Liman license in 2005.
- In the Aktobe area of central Kazakhstan, CNPC owns 100 percent of the onshore Zhanazhol and Kenkiyak fields, which together produced around 88,000 b/d in 2003. Output from these fields is targeted to reach 140,000 b/d.[23] In the Turgai basin of central Kazakhstan, CNPC in 2005 bought a Kazakh company producing about 4,500 b/d from the Arysskoye field, and another one producing about 8,000 b/d from the Kuatamlonmunai field.[24] CNPC also owns 60 percent of the onshore Uzen field in western Kazakhstan.
- In Kazakstan's offshore area, CNOOC and Sinopec tried, but failed, to buy into the Agip KCO Kashagan consortium when BG decided in 2003 to sell its ownership share. The would-be Chinese participants apparently held the view that being part of the development of Kashagan could help justify future investment in a west-to-east oil pipeline across Kazakhstan to China. The remaining partners exercised their preexisting rights to buy the full BG interest, but were forced to sell half the BG interest to KazMunaiGaz.
- Elsewhere in Kazakhstan's offshore, Sinopec acquired in 2004 four blocks producing about 4,000 b/d.
- In western Kazakhstan, ShengLi bought 50 percent of Big Sky Kazakhstan at the end of 2003. This first international purchase for ShengLi gives it access to the onshore Morskoe block, an old field drilled in the 1960s, 10 km from Tengiz, and to the Karatal and Dauletaly oil blocks about 300 km north of Morskoe.
- In mid-2004, CNPC signed an agreement with Uzbekneftegaz to boost cooperation, with an eye to future energy exports to western China. This agreement, confirmed in 2005 for about $600 million, calls for CNPC to explore with Uzbek partners. It underscores not only China's continuing drive for additional sources of oil but also its long-run hope to turn its western province of Xinjiang into a major production and transport hub.

Focus on the Caspian: Long-Distance Pipeline Proposals

Despite its possible interest in oil pipelines running through Pakistan, Bangladesh, and/or Myanmar, China seems to be pursuing a Kazakh oil pipeline line more aggressively than other alternatives, particularly after the flip-flops in Russian pipeline decision making. A Kazakh oil pipeline avoids the Malacca Straits and offers China a non–Middle East, long-term, and secure source of oil.

Over the years, however, proposals for building Caspian–Chinese pipelines typically generate high initial interest, are then shelved, and subsequently resurrected. Given China's growth in oil imports, Beijing's interest in future Kazakhstan crude supplies has grown. Kazakhstan, for its part, believes that if it can successfully seize a portion of the Chinese crude oil market, it gains a first-mover advantage for the future and negates or delays future export competition with Russia.

In mid-2004, one of these oil pipeline proposals was revived, and construction finished in 2005: the line from Atasu in central Kazakhstan to Alashankou on the Chinese border. It is expected to be in full operation in 2006, but it continues to suffer uncertain economics and throughput, as did its predecessors. Over the last 10 years, several Central Asia–China possibilities arose:

- In the mid-1990s, CNPC signed agreements with all the Central Asian countries to increase Sino-Central Asian oil cooperation.
- In 1994, CNPC and Turkmenistan signed a letter of intent to study a project for a natural gas pipeline to China. In 1995, CNPC, Exxon, and Mitsubishi agreed to build a gas pipeline from Turkmenistan through Uzbekistan, Kazakhstan, China and onward to South Korea and Japan.[25] The Central Asian countries were particularly interested in the project as a way to escape from the stranglehold of the existing Soviet-era natural gas pipelines, which were their only gas export routes, but the project came to naught. China thereafter completed construction of a $24-billion, domestic West-to-East natural gas pipeline running from the Tarim basin in the west to Shanghai and other demand centers in the east; this line might eventually carry future Central Asian gas supplies. That future appears distant, however, as China's own Tarim gas is already facing difficulty competing with cheaper coal in the eastern industrial and population centers.
- In 1995, Kazakhstan and CNOOC agreed on an oil pipeline that would link Kazakhstan and China's east coast. This was part of China's transboundary concept of a "Pan-Asian Continental Oil Bridge," including the possibility of a gas export pipeline to China from Russia's Siberian region.[26] By 1999 and in view of low oil prices at the time, CNPC

announced that reserves were insufficient in the Uzen and Aktyubinsk region for a 500,000-b/d pipeline project to be economic. The project did not completely die, but additional studies of a reduced capacity of 200,000 b/d further showed the project uneconomic. It was officially put to rest in 2003, a victim of insufficient reserves, costs (estimated at $3.5 billion), and doubts about financing.[27]

- In late 2003 a new variant of the above proposal emerged. An approximately 3,500-km Kazakhstan-to-China oil pipeline proposal surfaced, of which the 1,000-km construction of one segment, the $700-million line from Atasu in central Kazakhstan to the Chinese border city of Alashankou and nearby refineries, was completed in 2005. It is expected to be years before the whole line traversing Kazakhstan is completed, as negotiations continue between the two countries on contractual relations and prices. The Atasu–Alashankou project originally called for delivery of 800,000 b/d to 1.0 mmb/d, but was subsequently scaled back to an initial capacity level of 200,000 b/d (expandable to 400,000 b/d with additional pump stations), and estimated to cost $850 million.[28] This proposal, however, still suffers from many of the same weaknesses as did its predecessor: high construction costs and tariffs (estimated at $3.50 per barrel), which may make it uncompetitive; adequacy of reserves; and price competition with alternative sources delivered by ship to Chinese coastal cities.

  This oil pipeline across Kazakhstan will be built in stages, and is unlikely to be completed for at least 10 years. With the 2002 completion of the Kazakh lines from Aktobe to Kenkiyak and from Kenkiyak to Atyrau (448 km), oil from the Aktobe and Kenkiyak fields now moves south and west for export either by barge to Neka or through the CPC line, in the opposite direction from what is intended to reach China. This segment, however, is easily reversible for future eastward movement. A 700-km link between the Kumkol area (where PetroKazakhstan was active) and Kenkiyak will go onstream in the next year or so. Another section, of about the same length, from Kumkol to Atasu is under discussion, with construction tentatively aimed for sometime after 2011.[29] Uncertainties on reserves, costs, and financing remain, all of which plagued its predecessor protocols. Until completion of all the Kazakh sections, rising oil production from LUKoil and PetroKazakhstan (now CNPC), and of others in the Kumkol area, will move by rail west to Kenkiyak and then by pipeline westward, or by rail eastward to Atasu.[30]

Both Kazakhstan and China see advantages in constructing an oil pipeline linking the two countries, even if it were to be constructed on strategic terms rather than strict economic terms. Kazakhstan's interest derives from the flexibility such a route would give it. It could export oil from Atyrau west to the Black Sea on CPC, and from Atyrau east to China once all the sections of the proposed line are completed. On the one hand, China would be a monopsony buyer, just as in the Russian spur project, and once the pipeline were in place, could try to change the terms of purchase to Kazakhstan's (or in the former case, Russia's) disadvantage. China's interest in this proposal, on the other hand, stems from the cancellation of the Yukos proposal and some doubt as to whether Russia is committed to the Skovorodino–Daqing project. China is pushing the Kazakh scheme to gain oil import diversity and as a backup to compensate for further uncertainties surrounding future Russian exports. Nonetheless, the Kazakhstan–China line's construction and schedule remain highly uncertain.[31] Despite the plans, this line could be delayed or abandoned altogether, since Kazakhstan's oil export options remain wide open, and include shipping westward via the BTC line or southward through Iran.

Despite these oil export uncertainties, Kazakhstan is pushing, on a parallel track, a proposal to build a gas pipeline to tie into China's domestic West-to-East gas pipeline. Turkmenistan and Uzbekistan are likewise interested in future gas sales to China, and eventually to Japan and Korea.

Both phases of the West-to-East gas pipeline are complete and operational. Initially designed to carry 12 bcm/y, the pipeline could, if warranted, be expanded to 20 bcm/y by 2010.[32] The line is part of China's effort to increase gas consumption from its own basins, for both diversification and environmental reasons, and as a means of supporting economic growth. The target is to increase total gas use from the 2000 level of about 24 bcm/y to 100 bcm/y in 2010.[33]

The history of this gas project was not a pleasant one from the point of view of foreign investors. After more than 2 years of negotiations, PetroChina (formerly a part of CNPC) terminated discussions in mid-2004 on joint venture participation by Shell, ExxonMobil, and Gazprom. At one point BP had been a member of the foreign participants, but pulled out. China's former prime minister, Zhu Rongji, had pursued foreign participation and technology for large projects such as this one. With the departure of Mr. Zhu, PetroChina had little incentive to cooperate, and disagreements between it and the western companies grew. The ultimate stumbling block was the unwillingness of the Chinese to charge gas consumers in Shanghai and elsewhere high enough prices to provide for the target rates of return to the international investors. Higher gas prices would have made cheaper coal the preferred alternative, going against the government's desire to move away from coal to clean gas.[34] The pipeline was in effect financed by the state. According to the press, this failure of a foreign international company involvement was a great embarrassment to PetroChina, and signaled the possible future reluctance of foreigners to invest in China, if they expect that China will ultimately exclude their participation.[35]

In addition to the proposed new 30-bcm/y gas pipeline from Kazakhstan to China, estimated to cost $1.2 billion, gas pipeline alternatives include the proposed construction of

- An extension of the existing Kazakh gas system from Almaty to the Chinese border. Allegedly China prefers this route, since it is the least technically difficult, and could be built quickly.
- A pipeline from western Kazakhstan to Shymkent in the south, then onward to Almaty and the Chinese border.
- A pipeline from Omsk in the north to Astana in the south, and then onward to the Chinese border. Allegedly Kazakhstan prefers this route, since it would permit connecting with the Russian gas pipeline system as well as serving China.[36]

The disrepair, lack of safety, and expense of rehabilitation of the existing CAC gas pipelines serving Russia from Central Asia may make a new build the preferred solution. Kazakhstan, for example, has experienced four gas pipeline explosions on its system within the last 4 years, all of which testify to the age, corrosion, and leaks of the facilities. The Chinese market continues to beckon, however, and the West-to-East Chinese gas pipeline offers a potential link for Central Asian gas exports. Central Asian gas remains a long way from China's consuming centers, and pipelines will be costly to build and probably difficult to finance.

Much depends on how rapidly the Chinese gas market expands. This growth will determine whether the completed West-to-East pipeline will be expanded to and beyond 20 bcm/y and connected to Russian or Central Asian supplies, or both. To make either a Central Asian or Russian connecting line economic, potential sales would need to be 20–30 bcm/y to justify the investment; there are already doubts as to whether Turkmenistan has sufficient gas to meet its current contracts. Competition for the Chinese gas market is likely to intensify, however, between Central Asia and Russia, for China remains the nearest and largest potential gas buyer.

Conclusion

China's growing economy is likely to increase its energy demand in the years ahead, regardless of possible short-term hard landings. Kazakhstan could become a major overland exporter of oil to China, particularly if the two countries can cooperate on building the various pipeline segments, even though Nazarbayev has stated he prefers a line through Iran. Where the oil originates is yet to be determined; the line to China could be filled with a combination of oil from the central Kazakhstan Kumkol fields, future Kashagan and/or Tengiz output, and as-yet-undeveloped East Siberian

Russian oil as well. Connections of China's West-to-East gas pipeline to Central Asian natural gas supplies are not on the immediate horizon, but could follow if a completed oil pipeline demonstrates the success of Caspian energy trade with China.

UZBEKISTAN

With a population of more than 26 million, Uzbekistan is by far the most populous country of the Caspian region. Adjacent to Kazakhstan and Turkmenistan, Uzbekistan in our view is a member of the outer circle, since it does not border the Caspian Sea itself. Its energy significance for the region lies more in its pipeline network than in its gas or oil resources, as Soviet-era pipelines traverse its territory linking Uzbek supplies as well as those from Turkmenistan and Kazakhstan to Russia and westward. The country's major exports are gold and cotton, and it is far less dependent on energy development and sales than are its neighbors. At this point in time, its production levels are not relevant to the Caspian region's export future, and politically it is a nonplayer. Uzbekistan is currently self-sufficient in energy, but could become an energy exporter with sufficient investment capital and an appropriate investment climate. Other than the deal it recently concluded with China, Tashkent has no interest in attracting outside investment, and to all intents and purposes has told the West to stay away.

A Natural Gas Province Dependent on Russian Transport

Uzbekistan has several hydrocarbon basins and is primarily a natural gas region. The most productive basins historically were in the gas-prone Amudarya region across the border from Turkmenistan, and in the oil fields of the Fergana Valley. These are old fields dating from the Soviet era and need investment to boost production. Figure 4.1 shows the hydrocarbon basins that ignore the borders of Uzbekistan and Turkmenistan, and of Uzbekistan and Kazakhstan, while Figure 4.2 shows the oil fields in the Fergana Valley and the gas fields adjacent to Turkmenistan, together with the gas pipeline systems running northwest.

Uzbekistan's oil reserves are estimated at 0.6 billion barrels, only slightly above those of Turkmenistan, and well below those of Russia, Iran, Kazakhstan, and Azerbaijan. These are not sizeable enough to make the country a major contributor to world markets and to energy diversity and security. The country is at present a net oil and gas exporter, although not a major one. Self-sufficiency was achieved at a price, however, as the state oil and gas company, Uzbekneftegaz, is burdened with debt that it finds difficult to service.

While the country's gas reserves are somewhat smaller than those of Turkmenistan, they are still large at 1,870 bcm—greater than those of the

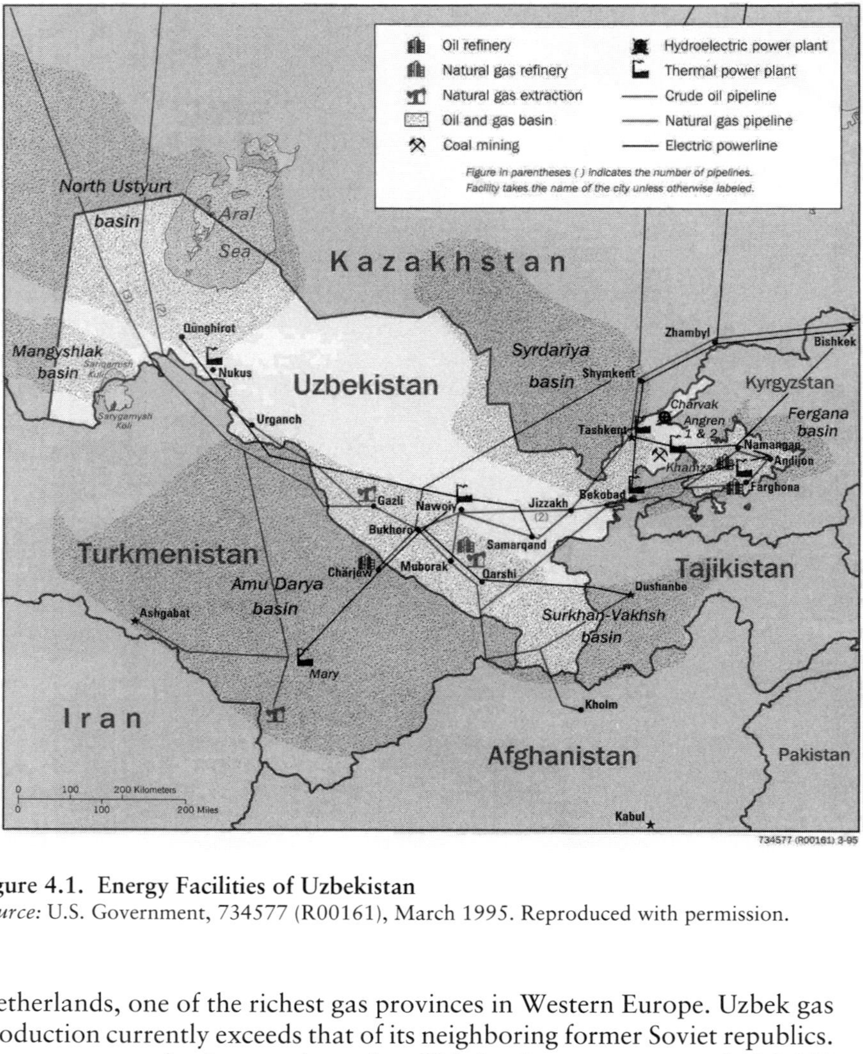

Figure 4.1. Energy Facilities of Uzbekistan
Source: U.S. Government, 734577 (R00161), March 1995. Reproduced with permission.

Netherlands, one of the richest gas provinces in Western Europe. Uzbek gas production currently exceeds that of its neighboring former Soviet republics. Most recent production numbers place Uzbek oil output at around 125,000 b/d and gas production at about 55.8 bcm/y.[40] Should Kazakhstan succeed in developing its own southern gas fields in the Amangeldy area, it will in the future cut back on gas imports from Uzbekistan.

Uzbekistan's production experience differed from that of other Caspian energy producers, for its output continued to rise during most of the post-Soviet period. Its oil production grew every year from 79,000 b/d in 1992 to 1999, when it peaked at 191,000 b/d. The country became a net oil exporter in 1996. Lacking foreign investment, its output has since declined, and Uzbekneftegaz expects it to fall even further. Gas production, however,

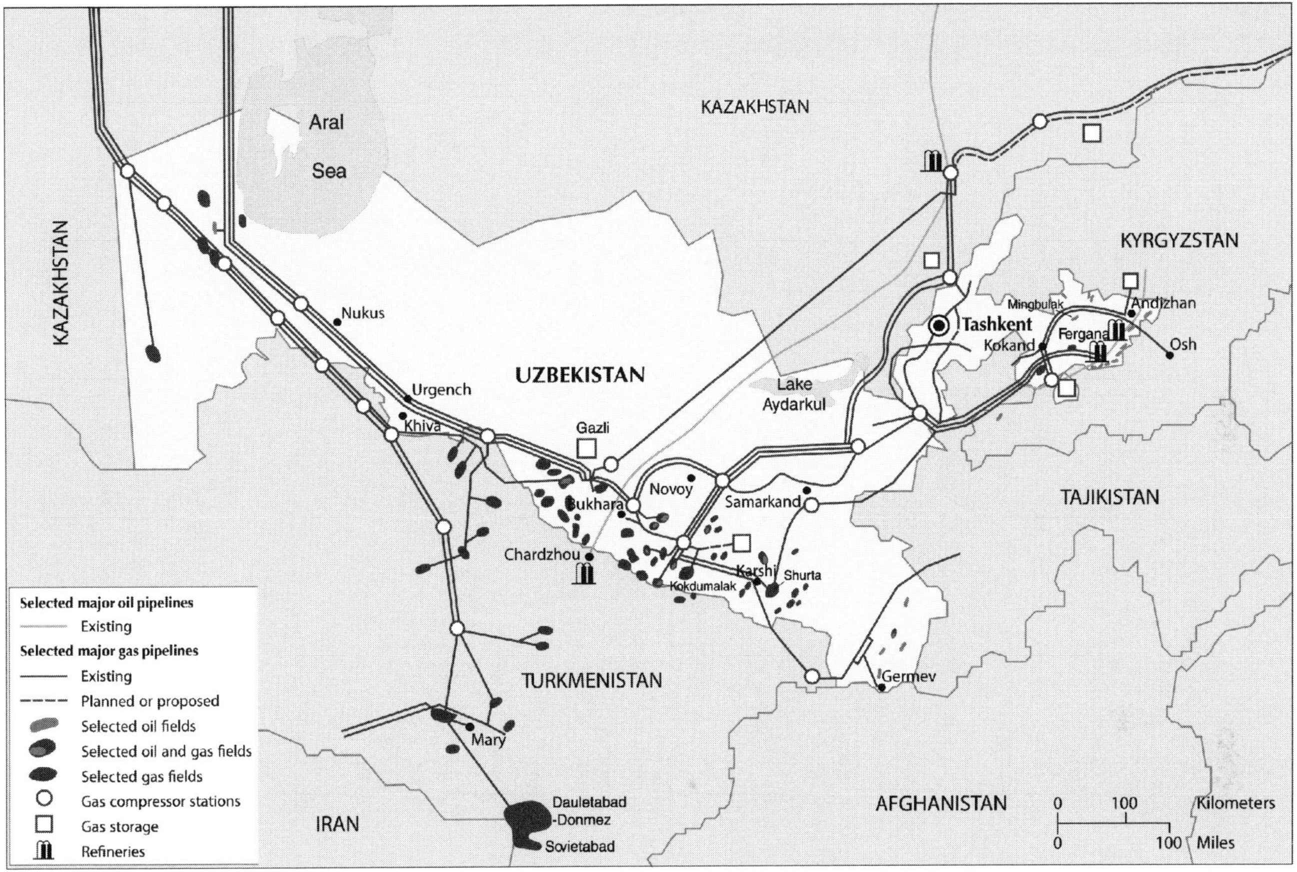

Figure 4.2. Natural Gas Export Pipelines of Uzbekistan, Turkmenistan and Kazakhstan
Source: International Energy Agency, *Caspian Oil and Gas: The Supply Potential of Central Asia and Transcaucasia*. Paris, France: OECD/IEA, 1998, 26. Reproduced with permission.

rose throughout the period, from about 40 bcm in 1992 to nearly 56 bcm in 2004. The country remained the largest gas producer in the area through 2004.[41]

While its natural gas reserves could permit Uzbekistan to become a major gas exporter, all its transport routes pass through Russia, the monopsony buyer. With Gazprom's expansion in 2004 of the capacity of the CAC I, II, and IV lines from 45 to 56 bcm/y,[42] Central Asian gas producers will be bound to Russia for some time to come as the sole route to move their gas, and they will compete with each other for customers.[43] The press suggests that Russia made an explicit choice in favor of spending $100 million on the CAC lines rather than billions of dollars on Yamal gas development.[44] Uzbekistan has no choice but to provide Gazprom with its exportable surplus of 10 bcm/y from 2004 through 2010, and perhaps as much as 20 bcm/y by 2020. This target will depend on greater pipeline capacity. Some analysts view the Russian contract, however, as a temporary Russian move to continue using cheap Turkmen and Uzbek gas until large volumes of less-distant and equally cheap Kazakh gas become available.[45] Russia allegedly prefers dealing with its adjacent Kazakh neighbor rather than with the unpredictable Niyazov, whose history as a supplier has been spotty, and/or with Karimov, both of whom Moscow (with some justification) regards as medieval Central Asian khans.[46] The problem for Uzbekistan, as also for Turkmenistan and Kazakhstan, is to find international gas markets and the routes to serve them without being held hostage to Russian pipelines. Until that occurs, Central Asian gas production will never reach its full gas potential and will remain largely stranded and unmonetized.

Little Foreign Interest

International interest in Uzbekistan's energy resource development, both upstream and downstream, is very low. Economic reform has moved extremely slowly, discouraging private foreign interest. Tashkent's policy to be left alone has repulsed foreign investment. Those few foreign companies in Uzbekistan have typically been service companies, such as Baker Hughes and China Petroleum Technology & Development, which have had contracts for upgrading Uzbek wells. Foreign internationals willing to perform exploration and development have not been in evidence. Until recently, the country's sole private PSA was with the UK's Trinity Energy, signed in 2001. Russia, however, is beginning to take an interest in Uzbek prospects, as LUKoil and the Uzbek government signed in mid-2004 a 35-year PSA for the Kandym gas development project (located close to the border with Kazakhstan). The LUKoil project had been under consideration since 2001. The PSA calls for production to begin in 2007, ultimately reaching a peak of 8.8 bcm/y, with gas sales to Gazprom or to Kazakhstan. The costs and timing of this project are uncertain, and in our view the production date will likely slip.[47] No other PSAs have been signed, although Gazprom and

Islam Karimov [AP Photo/Anvar Ikyasov]

Uzbekistan: Desperately Clinging to Power

Like the other leaders of the former Central Asian Soviet republics, Uzbekistan's Islam Karimov is a product of the Soviet years and has ruled his country as a dictator since before its independence. Following the May 2005 Andijan massacre, and the U.S. criticism of it, Karimov announced in late July 2005 the eviction of the United States from the Khanabad-Karshi airbase, giving the United States 180 days to depart. In our view, Uzbekistan is neither of energy or security significance to—nor a worthy ally of—the United States.[37]

Born in 1938 and raised, like Niyazov, in an orphanage, Karimov subsequently earned degrees as an engineer and economist from the Central Asian Polytechnic Institute and the Tashkent Institute of National Economy, respectively. He worked as an engineer at a Tashkent aviation complex and then moved in 1966 to the Uzbek state planning office. He was appointed in 1989 as Uzbek Communist Party first secretary. In March 1990, he was elected president of the Uzbek Soviet republic. Karimov declared Uzbekistan's independence, and won the rigged presidential election in December 1991. International observers cited extensive election fraud in that election. In 1995 a referendum was held to extend Karimov's term to 2000; he was reelected in a process that the U.S. State Department found "neither free nor fair." Karimov's current term expires in 2008. While he has stated that it will be his last, there is doubt whether he will honor that pledge, despite reports of his failing health. Some speculate he will arrange his appointment as president for life.

The Uzbek economy is largely closed and Karimov runs it as a centrally planned one. It is estimated to have declined more than 50 percent in the 3 years to 2005.[38] Karimov has not honored his repeated promises to the United States and others to make economic reforms and to improve the human rights situation in the country. He has forced closures of NGOs, continues to censor the media, torture and jail political opponents, and stage and countenance fraudulent elections, as the U.S. Department of State reports annually. The EBRD has dramatically reduced its lending to the country. Given the greater endowment of oil and gas resources elsewhere and Uzbekistan's restrictive investment climate and total state control of the energy sector, private foreign interest in exploration and production is low.

Uzbekistan appears to be at a political tipping point, on the verge of either explosion or implosion. Under Karimov it is spiraling downward with repeated cycles of violent unrest and repression. The leadership seems to be out of touch with reality, intent at all costs on keeping power in the hands of Karimov and his close supporters. His rule is fast losing all legitimacy as the economy has contracted and economic unrest has grown. The outcome is unlikely to be either bloodless or peaceful,[39] and will likely destabilize other fragile Central Asian regimes.

Uzbekneftegaz are expected to sign a second, 45-year PSA to develop gas fields in the western part of the country, near existing pipelines.[48] In mid-2005 China, however, agreed to provide $600 million for a joint venture between Uzbekneftegaz and CNPC for the development of oil and gas resources. Ironically, this agreement came on the heels of the suppression of the May 2005 Andijan unrest, with China supporting Tashkent's reaction to it.

Uzbekistan is another country where "the politics are not right" from the point of view of private international energy companies. The government is known for its state control and central planning, its repression, its absence of a predictable rule of law, and its corruption. The leadership announced that it would start restructuring Uzbekneftegaz in late 2004, but this has not occurred. The government had in earlier years announced restructurings but never followed up. While the 2004 proposal lacked details, only 49 percent of the company was to be sold off, thus guaranteeing the state a controlling interest in any future energy activities. The intent was to sell 39 percent to a "strategic foreign investor" and 10 percent to employees. A former plan to sell 49 percent of the company was cancelled because of absence of foreign investor interest. True to form, the 2004 privatization deal did not happen. The funds Uzbekneftegaz seeks may have to come from loans rather than foreign direct investment, thus making its debt burden problems even worse.

Foreign energy investors have better prospects elsewhere, both in geological terms and in terms of macroeconomic predictability and stability. Considerations working to deter Western interest in Uzbekistan include

- The fact that the Uzbek leadership really does not want Western companies or Western governments present
- The fact that the country is primarily a gas province rather than an oil one, in a region that already has a surfeit of gas[49]
- The twice-landlocked geographic position of the country, which makes access to international markets difficult through long and costly export routes
- The inconvertibility of the *som*, the national currency, until late 2003, coupled with other inefficiencies in the banking system, and the subsequent freezing of *som* convertibility in 2005
- The absence of clarity in the legal system and doubts about the sanctity of contracts
- The autocratic nature of the regime under President Islam Karimov
- The country's abysmal record of human rights abuses, and the loss of U.S. government confidence, evidenced by the State Department's decertification of the country as a recipient of continuing U.S. foreign aid
- Repeated clashes with Turkmenistan over rights to gas deposits near the border, and
- The threat of continued civil unrest and clashes in the Fergana Valley, where certain oil fields and facilities are located, and the risk that these clashes may spread.

Given that Uzbekistan has the largest population in the region, it will face in the future a higher level of internal gas and oil demand than will

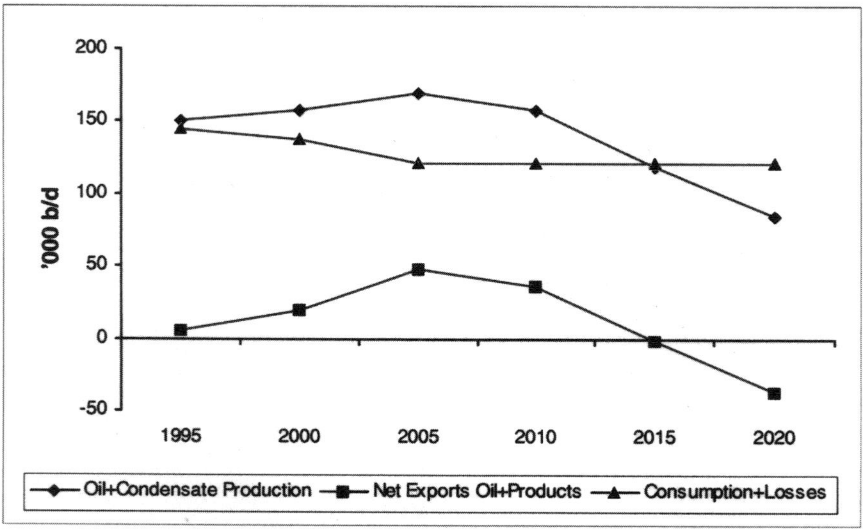

Figure 4.3. Uzbekistan's Oil Supply/Demand Balance to 2020 ('000 b/d)
Source: Wood Mackenzie, *Uzbekistan: Country Overview*, August 2003, 37. Reproduced with permission.

the others, unless of course the others develop faster and their industrial/commercial demand(s) more than outstrip the demand from the larger Uzbek population. According to Wood Mackenzie, the country is likely to lose its self-sufficiency status in both oil and gas around the year 2015, and become a net importer. Certainly the gas picture could be changed, given the size of the reserves, were exploration and development capital, not only from CNPC but from others as well, to be attracted. Figures 4.3 and 4.4 represent Wood Mackenzie's view of Uzbekistan's oil and gas production and export horizons to 2020.

Conclusion

Uzbekistan is not likely to have a major impact on world energy markets through its future exports of either oil or gas. Its energy significance lies solely in its pipeline routes and its links to neighboring countries. The country is left with the unappealing proposition of exporting its gas over the Russian gas pipeline system, or hoping that the resurrected proposal for the Trans-Afghan pipeline south to Pakistan and India goes forward, or wishing for a future gas line to China. Over the years various pipeline projects have been considered, but Uzbek gas outlets will be limited by neighboring countries' determination, such as Turkmenistan and Kazakhstan, to market their own excess gas. Until a new major gas export pipeline is developed, which in turn depends on the right investment and market climate, Uzbek and other

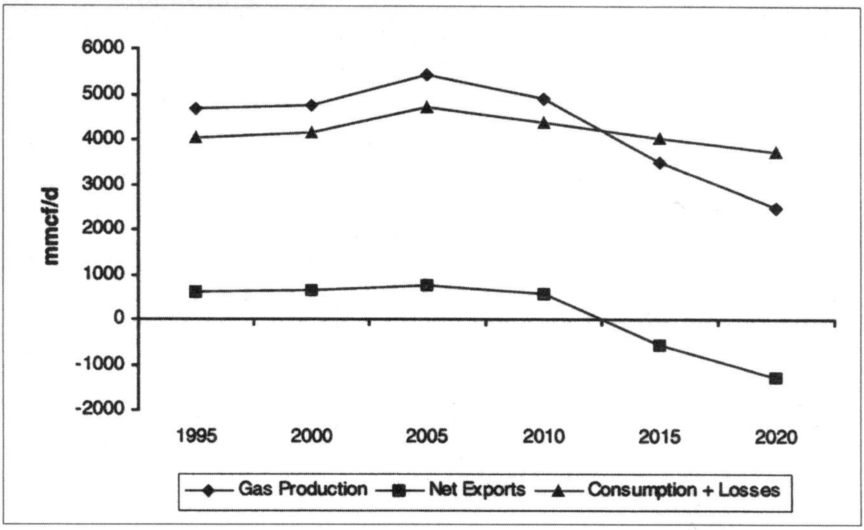

Figure 4.4. Uzbekistan's Natural Gas Supply/Demand Balance to 2020 (mmcf/d)
Note: The measure mmcf/d can be translated roughly into bcm/y by dividing the former by 100.
Source: Wood Mackenzie, *Uzbekistan: Country Overview*, August 2003, 39. Reproduced with permission.

Central Asian gas will flow either through Russia or will be left, stranded, in the ground.

With little energy significance, its political significance has suffered as well. The country's importance to the United States diminished with the collapse of the Taliban and the eviction of the United States from the Karshi–Khanabad airbase in mid-2005 (with 6 months' notice) in the aftermath of Andijan. Tashkent has proven to be an embarrassing U.S. ally, given its disastrous record on human rights and repression. As the country slides into further chaos and unrest, Uzbekistan has the potential to spur political instability in its neighbors and throughout the region.

With or without Karimov the despot at the helm, what role the United States will play in such an uncertain Uzbek future is far from clear. Karimov cares little for the citizens of his country; as a medieval khan, he cares only for himself and his continued position of power. He admits no need for change of any kind. His actions thus increase the likelihood of greater unrest by the poor and desperate and perhaps their cooperation with Islamic radicals. Should his regime collapse, only Islamic radicals look likely to have the cohesion to form a new government. The United States is in a difficult position as it pushes democratization but at the same time cannot fully distance itself because of the credibility of the Islamic movement. The United States needs to prioritize its interests, with democratic goals overriding the Department of Defense's interests in forward basing.

UNITED STATES

While it favored the adoption of democracy by the countries formerly a part of the Soviet Union, the United States gave little attention to the South Caucasus or to Central Asia both before and for several years after the demise of the Soviet Union. The isolation of Russia, however, was a goal of the Clinton administration from the outset. Regarding the other former Soviet states, the United States was on record as supporting democracy, free markets and trade, and human rights. It was primarily during the run-up to the signing of the AIOC contract with Azerbaijan in 1994 that Washington began to take an interest in what the Caspian might offer in terms of energy supplies. The region was and is seen as a new frontier not subject to the vagaries of OPEC pricing and output decisions and/or Middle East instability. The United States viewed it as an area whose development might improve world energy security through diversification away from Middle East supplies. U.S. attitudes toward the region changed, however, with the occurrence of the terrorist events of September 11, 2001. In this section we look at U.S. interests both before and after that date.

United States Interests in the Caspian Region, pre-9/11/2001

The United States has over the years attempted to develop an energy policy, but has been unwilling to take steps to raise taxes on oil so as to use price signals to restrain or cut oil demand. It has relied for the most part on diversification of its oil imports as an attempt to avert undue reliance on Middle Eastern sources of supply, given the frequent wars and political instability historically plaguing that part of the world. As the largest oil consumer in the world, the United States has long expressed interest in its companies' having access to overseas prospects, particularly those not controlled by state oil firms. President Clinton's 1997 national security strategy noted the potential for Caspian energy as playing an "increasingly important role in meeting rising world energy demand in the coming decades."[50]

The United States welcomed the opening up of the Caspian's oil and gas prospects to interested foreign investors after the fall of the Soviet Union. American companies were among the earliest to win contracts, primarily ChevronTexaco, Amoco (now part of BP), ExxonMobil, Unocal, Devon, and Hess. But the U.S. promotion of the BTC line was widely resisted and criticized, not only by Russia and Iran, who each wanted a part in the transportation of production, but also by academics and many of the companies themselves. Unremitting pressure from Washington together with changes in the views of the companies finally led to the decision to move forward with BTC, although doubts remain about its future economic success should Kazakhstan not provide sufficient throughput after the decline in ACG production.

The real question is why, as supporters of market-determined solutions, U.S. administrations pushed for what they considered a "strategic" pipeline at the expense of commercial considerations. This position included no intent of providing direct taxpayer support to assist in pipeline construction to underscore the "strategic" interest. It is well known that future growth markets for oil are in Asia, which means a pipeline south through Iran made sense, regardless of the concerns for security in the Straits of Hormuz, or the desire to help Turkey as a NATO ally. Unsaid for the most part were Robert Ebel's phrases "Anywhere but Iran" and "Anywhere but Russia," for routes through either would have been less costly than the route through Turkey. In fact, the United States looked hypocritical, urging Russia to move to a market economy and at the same time pushing for a "strategic" pipeline rather than a market-determined solution. By ruling out a pipeline in Iran, the United States forewent a chance to reach out and extend an olive branch toward improved relations with Tehran. By its subsequent promotion of the oil line in and through Russia to Murmansk, the United States pursued a policy in the later years that looked totally inconsistent with that of earlier years.

A number of authors writing in the last decade took issue with U.S. support of the BTC project. These authors also found very little U.S. strategic or energy interest in the Caspian region. They noted that the size of Caspian reserves and production would contribute little to world production and to overall world energy security. The countries are poor, militarily weak, thinly populated, and trade little with either each other or with the United States. These writers did not agree with the notion of a new "Great Game" for domination of the region, suggesting that for such a scenario there would have to be great interests at stake, and there are not. Certainly the United States was interested in new oil and gas supplies, but given the amounts involved, this interest was not an overriding one and not limited to the Caspian alone. These authors suggested that U.S. interest was at best derivative or tangential, in that it was determined by what intentions or actions larger powers in the region, like Turkey, Russia, Iran, or China, might exhibit.[51]

These views were consistent with the July 2000 Report of the Commission on America's National Interests, in which the Caspian is nowhere singled out as a region of national interest.[52] The distinguished members of this Commission went to great lengths to spell out what they considered were U.S. national interests in the years ahead. Moreover, they prioritized them in a hierarchy where the first group was defined as "vital," another "extremely important," a third "important," and the last of "secondary" importance. Vital interests would include those the United States is prepared to fight for, with or without international assistance, while extremely important ones might merit commitment of U.S. military forces to defend them, but only with international participation by others. Less important interests would merit a case-by-case assessment governing the use of U.S. military forces.

The Commission sets out the following five U.S. vital interests:

1. "Prevent, deter, and reduce the threat of nuclear, biological, and chemical weapons attacks on the United states or its military forces abroad;
2. Ensure US allies' survival and their active cooperation with the US in shaping an international system in which we can thrive;
3. Prevent the emergence of hostile major powers or failed states on US borders;
4. Ensure the viability and stability of major global systems (trade, financial markets, supplies of energy, and the environment); and
5. Establish productive relations, consistent with American national interests, with nations that could become strategic adversaries, China and Russia."[53]

Among these vital interests, only the fourth referenced energy—and most readers took this to mean not only oil and gas supplies, but also and in particular those supplies from the Middle East, rather than from less robust producing areas around the world. Other than the goal of keeping Kazakhstan nonnuclear after the return of its Soviet nuclear weapons to Russia, there were and are no vital U.S. interests in this area. At best, Caspian energy resources pale in comparison with the size of those in the Persian Gulf.

Moving down the hierarchy of interests, the Commission determined "extremely important" issues. These are those that if compromised would "prejudice but not strictly imperil" the ability of the U.S. government to safeguard and protect its citizenry. These are listed as to:

1. "Prevent, deter, and reduce the threat of the use of nuclear, biological, or chemical weapons anywhere;
2. Prevent the regional proliferation of WMD and delivery systems;
3. Promote the acceptance of international rules of law and mechanisms for resolving or managing disputes peacefully;
4. Prevent the emergence of a regional hegemon in important regions, especially the Persian Gulf;
5. Promote the well-being of US allies and friends and protect them from external aggression;
6. Promote democracy, prosperity, and stability in the Western Hemisphere;
7. Prevent, manage, and, if possible at reasonable cost, end major conflicts in important geographic regions;
8. Maintain a lead in key military-related and other strategic technologies, particularly information systems;
9. Prevent massive, uncontrolled immigration across US borders;

10. Suppress terrorism (especially state-sponsored terrorism), transnational crime, and drug trafficking; and
11. Prevent genocide."[54]

Nowhere in this second category would the Caspian region as an energy provider rank comfortably. Certainly the fourth item references any of the major Persian Gulf energy producers, such as Iran, Iraq, or Saudi Arabia, as being the target of prevention of emergence as a regional hegemon. The third and the seventh interests certainly have relevance for the Caspian region, where a number of rebellious or secessionist movements have taken up arms in attempts to either overthrow the existing state or to create a new state, but none of these would appear to imperil the U.S. government's ability to protect the nation or to merit commitment of U.S. forces. Should conflicts develop along proposed pipeline routes (item 7), it would not be fitting, in our view, for the United States to commit to protecting any or all future pipelines through the region in the interests of maintaining energy diversity. These are public-private projects that moved ahead after risks were assessed, and the parties to the projects are responsible for pipeline protection. With regard to the tenth item, all the countries under consideration have to date been firm in cracking down on either actual or perceived terrorist/radical movements within their borders. Crime and drug trafficking, however, still flourish and have not yet been adequately addressed by these new states.

Third in the hierarchy of interests, according to the Commission, are those termed "important," which—if compromised—would have "major negative consequences" for the ability of the U.S. government to protect and safeguard its citizenry in a free nation. These interests are to:

1. "Discourage massive human rights violations in foreign countries;
2. Promote pluralism, freedom, and democracy in strategically important states as much as is feasible without destabilization;
3. Prevent and, if possible at low cost, end conflicts in strategically less significant geographic regions;
4. Protect the lives and well-being of American citizen who are targeted or taken hostage by terrorist organizations;
5. Reduce the economic gap between rich and poor nations;
6. Prevent the nationalization of US-owned assets abroad;
7. Boost the domestic output of key strategic industries and sectors;
8. Maintain an edge in the international distribution of information to ensure that American values continue to positively influence the cultures of foreign nations;
9. Promote international environmental policies consistent with long-term ecological requirements; and
10. Maximize US GNP growth from international trade and investment."[55]

Items 2 and 3 avoid definition of what are "strategically important states" or "strategically less significant geographic regions." There are no major negative consequences to the United States in the event that Caspian energy resources should be slowed in their development, for supplies elsewhere in the world are on the rise. It is certainly true that Caspian resources will ultimately be marketed on either a global or regional basis, and will diversify—as long as they last—the world's energy supplies. They will slow—but not eliminate—the path toward growing future dependence on the Middle East.

Finally, the Commission defines less important/secondary U.S. national interests as desirable conditions, but ones that really do not affect the U.S. government's capacity to protect and safeguard its citizenry. These include

1. "Balancing bilateral trade deficits;
2. Enlarging democracy everywhere for its own sake;
3. Preserving the territorial integrity or particular political constitution of other states everywhere; and
4. Enhancing exports of specific economic sectors."[56]

Since independence from the former Soviet Union, the United States, regardless of administration, has urged these Caspian states to move toward democratic government. That this desired outcome has not been reached is to admit to the difficulty of the transition process, or to admit that while the West favors democracy, the countries/leaderships do not. Leaders in the region have concentrated on maintaining stability through their own accumulation of power. If anything, democratic tendencies have been stifled and suppressed in the interests of the current generation of leaders' desires to preserve that power. No one knows the probabilities of democratic tendencies emerging after the departure—by age or by coup—of the current leadership.

In summary, nowhere in this array of hierarchical interests do the Caspian states and their energy resources fit well. Caspian energy is not a vital U.S. interest, nor is it either an extremely important or only important interest. We conclude that a diversified array of crude oil exporters is in the interests of the United States, and that the Caspian region is just one of many that are or soon will be providing the world with new energy supplies.

United States Interests in the Caspian Region, post-9/11/2001

Many things changed on September 11, 2001. The United States recognized the importance of this region for its potential contribution to the war against the Taliban and the broader global war on terrorism. Certainly the events of September 11 demonstrated as well the need for diversification of energy supplies. While the United States remains committed in principle to strengthening democracy and prosperity of the newly independent states, and preserving their sovereignty, the difficulty for it, both before and after

9/11, was and is in trying to advance along these fronts. In particular, the goal of strengthening prosperity and democracy has not been able to coexist with the goal of enhanced global energy security, since the energy resources lie in countries that have become increasingly undemocratic and authoritarian over the years since independence. In large part, Caspian leaders, while giving lip service to democracy, have stated that their countries are not ready for it, and they rule as dictators.

But even as the United States welcomes any new oil and gas reserves and production as contributing to strengthening the world's energy security, its post-9/11 interests in the Caspian region have changed dramatically. The United States now sees these countries as strategically significant not so much for future energy sources, but as key political and geographic allies in the war on terrorism. Post 9/11, the war on terrorism for the most part meant that military and security assistance to these countries ranked high on Washington's agenda, and trumped both energy interests and democracy building.[57] U.S. forces are now stationed in Georgia and Kyrgyzstan, and, until recently, in Uzbekistan.

From the current U.S. perspective, this is a region where the overriding national interest is to combat terrorism. No longer are the primary interests democracy, prosperity, or energy. Suppressing terrorism is one of the Commission's "extremely important" issues, but now in practice it seems to be elevated to a vital one. The shift in policy direction was evidenced by the United States' relationship with Uzbekistan and the use of its Karshi-Khanabad airbase to pursue the Taliban in Afghanistan as well as to be prepared to use force elsewhere in the region. In 2005, Uzbekistan ordered the United States to leave the airbase in retaliation for U.S. criticism of the uprising and massacre in Andijan. It is important to note that the United States has just about no other ties to Uzbekistan, and its departure within the 6-month period granted by Uzbek officials signaled abandonment of U.S. goals of advancing democracy there. The United States now has the opportunity, however, to reverse its pursuit of stability at the expense of democracy, as Secretary of State Rice noted in a June 2005 speech. The concentration to date on combating terrorism has led the United States to overlook some of its other interests, like discouraging massive violations of human rights and building or enlarging democracy, both goals of the Commission on America's National Interests.

The United States needs to ask itself if it is supporting regimes that do not meet our stated standards of openness and governance. The regional autocrats, particularly on the eastern side of the Caspian, use the threat of radicalism or terrorism as an excuse for denying human rights and diminishing the few freedoms that their citizens might have enjoyed. The State Department's annual reports on the human rights situations in these countries paint a very gloomy picture. The leaders in these countries justify their executive concentration of power, their sham elections, the curtailment of liberties, and the absence of participatory government, all in the name of national security against the war on terrorism or in favor of preserving stability. The United States, while recognizing that some of the leaders pose a threat of embarrassment to Washington, continues to support the leadership and has played along.[58]

Conclusion

The Commission on America's National Interests prioritized American interests in the pre-9/11 world. These interests still hold more than a half decade later. Many would argue that the prioritization also remains correct in a post-9/11 world. Caspian oil and gas development and production, while of appeal as a small net addition to world energy supplies, is not a vital or highly important U.S. national interest.

Washington's policy toward the region remains muddled. The former official interest in Caspian energy supplies as a means of building participatory democracy, safeguarding human rights, and improving living standards has been replaced with the war on terrorism. The United States has made little progress in advancing human rights in any of these countries, and has been evicted from the Uzbekistan airbase as a result of U.S. criticism of Karimov's position on human rights. To the extent that the United States is perceived as close to, and supportive, of the rule of dictators or repressive leaders, it risks perception in these countries as a hypocrite that advances American democratic principles and values but in fact supports tyrants and despots. The face of authoritarianism, seen as aided and abetted by the United States, augurs a fertile environment for the growth of resistance and of terrorism in these states. U.S. support of these repressive regimes risks an outcome fully opposite to the declared one of democracy and prosperity.

In addition, the U.S. position of support for repressive regimes risks the perception by both the Caspian leaders and their populaces that the United States is interested in oil and gas development at any price. Caspian leaders will try to use U.S. doubts about Russia's and other countries' energy production prospects to highlight the role the Caspian might play in enhancing U.S. energy security, but this rings false since it is a world market. Nonetheless, Caspian politicians have failed to note that their perception of U.S. need for oil at any price is not consistent with the continuing U.S. sanctions on Iran, and with its historical sanctions on Libya; both are examples of U.S. closure of avenues to increasing and diversifying energy supplies.

This author concludes that both before and after the events of September 11, 2001, the prospects for Caspian energy developments were overhyped and exaggerated. We conclude that it was a mistake in foreign and energy policy for the United States to have entered the energy fray in pushing for the BTC and CPC pipeline routes, and the Murmansk route. By so doing, the United States directly contravened its stance that free markets work best in providing market solutions. The United States has also been shown to be hypocritical in its support of foreign despots despite its alleged commitment to transparency and to human rights and freedoms. While Secretary Rice's remarks may signal a change going forward, both democratic goals and energy interests have been subservient to those of terrorism prevention. This policy is likely to have severe repercussions towards destabilizing the whole region in the future, should resistance to the local Caspian and Central Asian governments take to the streets in force. The leadership in most of these countries is working to further consolidate its power and suppress opposition. When the populations of these countries revolt against the leadership, the resulting instability, turmoil, and violence will be blamed on the United States.

THE OUTER CIRCLE: CONCLUSION

China desires new oil and gas supplies from wherever it can secure them. In its acute disappointment at the apparent cancellation of the pipeline proposal to carry Russian oil to Daqing, followed by Russia's subsequent announcement that it may yet build this line, Beijing is looking elsewhere, including at both Iran and the Caspian, to line up future supplies of oil and perhaps eventually of gas as well.

For more than 15 years, Uzbekistan has been an irrelevant player on international energy (primarily natural gas) markets. It remains captive to the old Soviet gas pipeline system for exports. Given its decisions to reject Western interest, clamp down on its own population, and evict the U.S. military, only China appears to remain interested in potential Uzbek oil and gas resources.

This author's judgment is that the Caspian states have yet to be persuasively defined as strategically important states for energy reasons to the United States. The United States is left with no vital or important energy interests in Caspian oil and gas supplies, and should not consider it necessary to risk lives or use force to protect energy resources there. As policy interest shifted from energy security to combating terrorism, the United States is increasingly seen as a voracious energy consumer who supports repressive regimes in order to add to oil and gas production. The United States has spent much political capital in pushing for an oil export route to the west through Turkey, even though the growth markets are in China and elsewhere in Asia. Future U.S. administrations would be wise to remove themselves from active intervention in commercial decisions on oil and gas development and export projects.

5

Concluding Remarks

Caspian energy developments have been the subject of much hope and wishful thinking. The reserves are not now thought to be anywhere near as great as those reported and circulated in the 1990s. All told, production from them is expected to contribute only a small percentage of future world oil demand, in the neighborhood of 3–3.5 percent. According to the priorities set out in the 2000 report of the Commission on America's National Interests, this is not a region of much strategic national energy interest for the United States.

Oil production in these countries is both landlocked and expensive. Markets are distant, and geologic and environmental challenges great. Gas production is constrained by the absence of developed markets and the absence of pipelines other than those in place from Soviet days. A number of fields have either been termed uncommercial, or are largely gas and thus not as desireable for developing and marketing as oil fields would be. Timetables slip and production levels are revised downward, below the estimates of the 1990s, in the face of both technical and political challenges. We maintain that Caspian production is costly, it will be less than what many people have predicted, and that it will likewise arrive on markets later than projected.

We have little hope for the future political or economic liberalization, or the adoption and practice of true democracy in the Caspian countries. They have neither the intent nor the capacity to hold true democratic elections. We expect them to continue with corrupt and autocratic leaders who repress the populations. While the likelihood of an imminent Rose or Orange revolution is probably low, economic despair is growing in the region. This despair feeds malcontents, who will look for ways to oust the dictatorships in Uzbekistan and Turkmenistan, and perhaps even in Kazakhstan. These countries are generally potential powder kegs waiting to explode, as corruption and the spoils from oil work against the common welfare or any hope for democracy. The war on terror has provided a convenient excuse to engage in additional capriciousness and repression across the region. Human rights exhortations have come to naught. Energy firms have already written off Turkmenistan and Uzbekistan at least for now for political reasons, regardless of the known energy reserves located there. Georgia is key in energy transportation for oil and gas, but it remains a thoroughly corrupt and fragile state; the pipelines offer tempting targets for disruption for whatever the "cause" or grievance, and we doubt that either Georgia or the companies will be able to prevent interruptions in energy flows. While Kazakhstan and Azerbaijan have kept Western investor interest, despite the worsening investment climate in the former, these countries are at risk of Dutch Disease as well as civil unrest, should they fail in diversifying their economies beyond the energy sector. Neither of these latter two countries can in any way be considered even a nascent democracy, as arbitrary arrests and beatings continue and the rule of law under an independent judiciary is nonexistent.

Turkey, Russia, Iran, and China have clear interests in Caspian energy production. Turkey is the link to connecting Caspian supplies with Western markets without going through the Turkish Straits, but its own demand, particularly for gas, has been vastly overestimated. Iran, Russia, Azerbaijan, and Turkmenistan are all competing for shares of the Turkish natural gas market as well as onward sales to Europe, if and when the gas pipeline network expands to the west.

The politics of energy development are often difficult, and this continues to be the case in the Caspian region. Corruption is rampant and either drives away potential investors or raises costs for current investors. Some authors have maintained that the State Department's use of the 200 billion barrels of oil reserves served as a tool for Caspian leaders to try to attract foreign investment. While foreign monies have been invested, the caution of developers is well known, particularly in the face of a deteriorating investment climate and the demands of the states' leaders. Some companies have already pulled out, especially in Azerbaijan, and BG has left the Kashagan project. Iran and Turkmenistan have not agreed to Caspian Sea delimitation, and Russia pressures CPC and waffles on future oil and gas sales to China and Japan. Iran, Russia, and the other Caspian countries have also been unable to agree on addressing the multitude of environmental problems that exist both offshore and onshore, some due to the energy industry but others due to the polluted industrial legacy of the Soviet era.

Much has been written on pipeline politics. Despite this author's disagreement with the U.S. government's insistence on the construction of the BTC oil pipeline instead of any of the other competing projects, this pipeline is completed and began loading Caspian oil on tankers in Ceyhan in 2006. The discussion has now turned to what other pipelines might be required to carry Caspian region oil and gas to future markets. In this author's view, there will indeed be a Bosporus bypass to avoid the risk of accidents, congestion and weather delays of the Turkish Straits, but there will not be a dedicated oil export pipeline for the anticipated production from Kazakhstan's Kashagan oil field. The BTC line, together with the CPC line—both of which can be expanded—should be sufficient, along with the growing volumes swapped at northern Iranian ports. Whether a long-distance oil or gas pipeline will be built to move Central Asian oil or gas to Chinese markets is a more difficult prediction, and may ultimately be a political one. At present, an oil pipeline from Atasu in central Kazakhstan to China is complete, although both the reserves to fill it and its costs are uncertain. In any event, the United States should not intervene in future choices for energy transportation, but follow its oft-stated principles and let the market determine the outcomes.

The Caspian region, both to the east and west of the Sea, is an area marked by authoritarian control and desperate poverty. It has no tradition of democracy; democracy has yet to even begin to take root in a meaningful way. The issue that may be far more telling than the richness of the hydrocarbon resources could be whether these nations stand at the brink of becoming budding democracies, or whether they will continue to succumb to strong-man rule, corruption, no rights of expression, and no participation in public affairs. Our judgment is that they will follow the latter route. The countries' leaders, possibly with the exception of Georgia, are thoroughly corrupt and the name of the game is to maintain power, fleece the state, buy favors, and line pockets. The region has a number of simmering conflicts of a military, ethnic, or religious nature that could catch fire once again. The area's very instability makes it an extremely risky area in which to extract and transport hydrocarbons, as the rules are always subject to change with *ex post* interpretations. Those who enter the exploration, production, and transportation fray clearly expect the rewards to compensate for the risks.

This region is one of many hydrocarbon provinces in the world. To overstate its importance, both in terms of its resources and its political and economic potential for development and contribution to world energy security, would be a serious mistake. The United States should recognize the political realities of the region, the likelihood for continuing instability if not total chaos, and temper its misplaced enthusiasm for the energy, economic, and political prospects of the region.

Notes

CHAPTER 1

1. Steven R. Mann, "Caspian Futures," in *Russian-Eurasian Renaissance?* edited by Jan H. Kalicki and Eugene K. Lawson, 147–160 (Washington, DC, and Stanford, CA: Woodrow Wilson Center Press and Stanford Business Books, 2003).

CHAPTER 2

1. For an excellent historical review of Caspian gas resources, see Ottar Skagen, *Caspian Gas* (London, UK: Royal Institute of International Affairs, 1997).

2. Fiona Hill, "Pipeline dreams in the Caucasus," *The Caucasus and the Caspian: 1996–1997 Seminar Series,* Harvard University, John F. Kennedy School of Government, Cambridge, MA, September 1996, available online at http://www.ciaonet.org/conf/jfk01/index.html.

3. AIOC's membership has changed since the original agreement. It presently consists of BP with 34.1% (after acquiring Amoco and its interest), Unocal 10.3%, SOCAR 10%, Inpex 10%, Statoil 8.6%, ExxonMobil 8.0%, TPAO 6.8%, Devon 5.6%, Itochu 3.9%, and Delta Hess 2.7%.

4. Tengizchevroil is a joint venture partnership of ChevronTexaco (50 percent), ExxonMobil (25 percent), KazMunaiGaz (20 percent), and LUKArco (5 percent).

5. Anonymous, "Worldwide Reserves," *Oil & Gas Journal*, Vol. 94, Iss. 53, December 30, 1996, 40 (2), gives reserves as of January 1, 1997. Since that time, reserve estimates for the Middle East have grown to 743 billion barrels, as shown in Anonymous, "Worldwide look at reserves and production," *Oil & Gas Journal*, Vol. 103, Iss. 47, December 19, 2005: 25.

6. Rosemarie Forsythe, *The Politics of Oil in the Caucasus and Central Asia*, Adelphi Paper No. 300, International Institute for Strategic Studies (London: Oxford University Press, 1996), 6.

7. *Ibid.*, 7.

8. *Ibid.*, 11.

9. U.S. Department of State, Barbara Larkin, "Caspian region energy development report as required by HR 3610," Congressional Record, May 8, 1997, S4208. This report, as directed by Congress, was prepared by the U.S. Department of State in consultation with the Secretaries of Energy and the Treasury, the Agency for International Development, the Export-Import Bank, Overseas Private Investment Corporation, and the Trade and Development Agency.

10. *Ibid.*, 2.

11. See Leonard L. Coburn, hearing before the U.S. Senate Foreign Relations Committee, on "United States energy security: Russia and the Caspian," April 30, 2003, available online at http://foreign.senate.gov/testimony/2003/CoburnTestimony030430.pdf.

12. Dilip Hiro, "Why is the US inflating Caspian oil reserves?" *Middle East International*, September 12, 1997, 19.

13. BP, *Statistical Review of World Energy*, June 2005.

14. Anonymous, "Worldwide look at reserves and production," December 19, 2005, *op. cit.*, 23–24. *OGJ* gas reserves reported in tcf and changed to tcm by author.

15. BP, *Statistical Review of World Energy*, June 2004.

16. Anonymous, "Worldwide look at reserves and production," *op. cit.*, December 20, 2004, 22–23. *OGJ* gas reserves reported in tcf and changed to tcm by author.

17. Anonymous, DOE/EIA, *Country Analysis Brief*, "Caspian Sea region: Key oil and gas statistics," August 2003, available online at http://www.eia.doe.gov/emeu/cabs/caspstats.html.

18. DOE/EIA, *International Energy Outlook 1998*, 14.

19. International Energy Agency, *Caspian Oil and Gas: The Supply Potential of Central Asia and Transcaucasia* (Paris: 1998), p. 32 for reserves for the region, and individual country chapters reporting reserves, not necessarily proved reserves, available online at http://library.iea.org/dbtw-wpd/textbase/nppdf/free/1998/caspian_oil_gas.pdf.

20. Yelena Kalyuzhnova, "Economics and energy," in Yelena Kalyuzhnova, Amy Myers Jaffe, Dov Lynch and Robin C. Sickles, eds., *Energy in the Caspian Region: Present and Future* (Hampshire, UK, and New York, NY: Palgrave Publishers Ltd., 2002), 58, hereafter called Kalyuzhnova, *et. al.*

21. Alec Rasizade, "Azerbaijan after a decade of independence: Less oil, more graft and poverty," *Central Asian Survey*, Vol. 21, Iss. 4, 2002: 350, available online at http://proquest.umi.com/pqdweb?index=12&did=776862991&SrchMode=1&sid=1&Fmt=2&VInst=PROD&VType=PQD&RQT=309&VName=PQD&TS=1129559037&clientId=3921.

22. Laurent Ruseckas, "Caspian oil development: An overview," in *Caspian Energy Resources: Implications for the Arab Gulf* (London, UK, and New York, NY: I.B. Tauris & Co Ltd. for The Emirates Center for Strategic Studies and Research, 2000), 15, hereafter called Emirates Center 2000.

23. Paul R. Gregory, "Developing Caspian energy reserves: The legal environment," in Emirates Center 2000, 25.

24. Geoffrey Kemp, "Iran and Caspian Energy: Prospects for cooperation and conflict," in Emirates Center 2000, 55.

25. Wilfrid L. Kohl, "The development of Caspian Sea oil: Implications for OPEC," in Emirates Center 2000, 127.

26. Michelle Michot Foss, S. Gürcan Gülen, and Bhamy Shenoy, "Caspian pipeline prospects hinge on transparent, logical regulatory frameworks," *Oil & Gas Journal*, Vol. 98, Iss. 34. August 21, 2000: 68 (5).

27. Amy Myers Jaffe, "Price vs. market share for the Arab gulf oil producers: Do Caspian oil reserves tilt the balance?" in Emirates Center 2000, 150.

28. Sarah A. Emerson, "The relevance of Caspian Oil for the world market," in Emirates Center 2000, 173, 186.

29. Friedmann Muller, "Policy options in the Caspian Sea region," in Dieter Dettke, ed., *A Great Game No More: Oil, Gas and Stability in the Caspian Sea Region* (Washington, DC, and Berlin, Germany: Friedrich-Ebert-Stifting, 1999), 8, hereafter called Dettke.

30. Laurent Ruseckas, "The Caspian energy resources and the world's energy needs," in Dettke, *op. cit.*, 19.

31. John Lichtblau, "U.S. policy and action toward Central Asia and the Caspian Sea countries," in Dettke, *op. cit.*, 68.

32. Patrick Love, "The changing face of energy politics," *The OECD Observer*, Summer 1999, No. 217/218, 2, available online at http://www.ciaonet.org/olj/oo/oo_99lop02.html.

33. Robert Ebel, "Caspian oil: Separating myth and reality," in *Central Asia and the Caucasus: A New Version of The Great Game. The Findings of a Conference Organized by the DACOR Bacon House Foundation* (Washington, DC: DACOR Bacon House Foundation, 1998), 73.

34. DOE/EIA, *Country Analysis Brief*, "Kazakhstan," November 2004, available online at http://www.eia.doe.gov/emeu/cabs/kazak.html.

35. See *BP Statistical Review of World Energy*, June 2005, 4; see also Anonymous, "Worldwide Look at Reserves and Production," *Oil & Gas Journal*, Vol. 103, Iss. 47, December 19, 2005: 25. The EIA does not publish an independent series on world oil reserves.

36. Kazakhstan is the exception: it has not only oil and gas, but also diamonds, other minerals and metals, and agricultural products.

37. Production numbers provided by DOE/EIA. Percentages calculated using *OGJ*, December 19, 2005, *op. cit.*

38. The CPC is multinationally owned, with the ownership shares as follows: Russian Federation—24 percent; the Republic of Kazakhstan—19 percent; the Sultanate of Oman—7 percent; ChevronTexaco—15 percent; LUKArco BV—12.5 percent; Rosneft/Shell Caspian Ventures Ltd—7.5 percent; Mobil Caspian Pipeline Company—7.5 percent; Agip International (NA) NV—2 percent; BG Overseas Holding Ltd.—2 percent; Kazakhstan Pipeline Ventures LLC—1.75 percent; and Oryx Caspian Pipeline LLC—1.75 percent.

39. See, for example, Anonymous, "Oil giants doubt bright Kazakhstani plans," *RosBusinessConsulting—News Online*, October 9, 2003, available online at http://www.rbcnews.com/free/20031009154509.shtml.

40. Email correspondence with industry expert, June 24, 2004.

41. See Mamdouh G. Salameh, "Middle East oil: What's the alternative?" United States Association for Energy Economics, *Dialogue*, Vol. 11, No. 3, Table 2, 16, November 2003.

42. See BP, *Statistical Review of World Energy*, June 2003 and June 2005.

43. Mark McCafferty and Valentina Kretzschmar, "CPC, BTC pipelines make current Caspian area oil export capacity adequate," *Oil & Gas Journal*, Vol. 102, Iss. 18, May 10, 2004, 52 (7). Figure 10 in this article suggests that domestic demand in Azerbaijan and Kazakhstan is unlikely to reach 400,000 b/d. Our estimate allows for greater economic growth and thus greater demand.

44. Oils are measured by the standards of the American Petroleum Institute (API) and are classified by gravity. Light oils with high-degree gravity are generally more valuable in the market than heavier oils.

45. Email correspondence with industry expert, June 24, 2004.

46. For a thorough discussion of the concept of energy security, see J. Bielecki, "Energy security: Is the wolf at the door?" Special Issue: Oil and the Economy: Recent Developments in Historical Perspective, *The Quarterly Review of Economics and Finance*, Vol. 42, Iss. 2, 2002: 235–250. See also Lord John Browne, Group Chief Executive, BP, *Energy—the Medium-Term Outlook* (Washington, DC: Washington Press Club, March 23, 2004).

47. DOE/EIA, *Annual Energy Outlook 2004*, January 2004, Table A21, 160, and DOE/EIA, *Annual Energy Outlook*, February 2005, Table A20, the latter available online at http://www.eia.doe.gov/oiaf/aeo/pdf/appa.pdf, 28–29; Sam Fletcher, "World oil production to rise by 16 million b/d by 2010," *Oil & Gas Journal*, Vol. 103, Iss. 7, February 21, 2005, 32–33, available online at http://ogj.pennnet.com/Articles/Article_Display.cfm?Section=ARCHI&ARTICLE_ID=221680&VERSION_NUM=1&p=7. Ken Chew of IHS Energy Group is the source of the 13 mmb/d figure (see slide 43 of his March 2, 2005, presentation entitled "Global oil supply issues: Recent trends and future possibilities," available online at http://www.ihsenergy.com/news/presentations/appex_2005/chew.pdf).

48. A representative of ChevronTexaco indicated that until the CPC pipeline opened in 2001, the company was the largest owner of railway freight cars in the Caspian region; all were used for transporting oil.

49. The literature is extensive on pipeline proposals. It includes, among others, Gawdat Baghat, *American Oil Diplomacy in the Persian Gulf and the Caspian Sea* (Gainesville, FL: University of Florida Press, 2003); Sander Hansen, *Pipeline Politics: The Struggle for control of the Eurasian energy resources* (The Hague, Netherlands: The Clingendael Institute, April 2003), 1 (77), available online at http://www.clingendael.nl/publications/2003/20030400_ciep_paper_hansen.pdf; Emmanuel Karagiannis, *Energy and Security in the Caucasus* (London, UK: Rutledge-Curzon, 2002); Hooman Peimani, *The Caspian Pipeline Dilemma: Political Games and Economic Losses* (Westport, CT: Praeger Publishers, 2001); Shams-Ud Din, ed., *Geopolitics and Energy Resources in Central Asia and Caspian Sea Region* (New Delhi, India: Lancer's Books, 2000); and Charles van der Leeuw, *Oil and Gas in the Caucasus & Caspian: A History* (New York, NY: St. Martin's Press, 2000).

50. This discussion of pipelines draws on Thomas R. Stauffer, "Caspian fantasy: The economics of political pipelines," *The Brown Journal of World Affairs*, Vol. VII, Iss. 2, Summer/Fall 2000: 63–78.

51. The economics of pipelines explains why most are built to an initial capacity level, with the option to expand capacity at a later date if circumstances warrant. This staged construction avoids excess capacity dragging down profitability.

52. Stauffer, *op. cit.*, 64.

53. Discussion with industry executives, October 21, 2003.

54. Robert Smith, "Politics, production levels to determine Caspian area energy export options," *Oil & Gas Journal*, Vol. 99, Iss. 22, May 28, 2001: 33 (6). This author uses $18, as does Fadhil J. Chalabi in his "Gulf oil vs. the oil of the Caspian Sea," in Emirates Center, while Mamdouh G. Salameh, in his "Caspian Sea is no Middle East," *International Association for Energy Economics Newsletter*, Third Quarter 2002, 12, uses $12–15 per barrel.

55. Gas marketing is very different from oil marketing, for oil is a fungible product. Once the oil is on a tanker, it can be delivered to the most attractive market. But gas pipelines, once constructed, are in place for up to 40 years or more, and are committed for the most part to long-term contractual agreements between producer and customer.

56. "Early oil" means oil exported from Azerbaijan by AIOC prior to determination of the ultimate Main Export Route, which is now the BTC route, and the construction of that line.

57. The right of way exists, and capacity can be expanded rapidly and cheaply through adding more powerful pumps or more pumping stations as well as looping the line (adding pipe parallel to the existing line).

58. See The Economist Intelligence Unit, "Oil export routes: The Baku-Tbilisi-Ceyhan (BTC) oil pipeline," *Country Report on Georgia*, August 2005, 21, citing BP on an additional 350,000 b/d through DRAs, and citing Michael Townshend, head of BTC, on an additional 300,000–400,000 b/d through looping and additional pump stations.

59. From a political perspective, AIOC kept interested parties happy, as it sent some oil north through Russia and some west through independent Georgia. AIOC is believed to have bowed to intense pressure from the United States to send its major ACG volumes through the line to Turkey.

60. Author's conversations with an AIOC representative indicated that the company decided that it was getting nowhere objecting to BTC in the face of U.S. pressure—"just banging our heads into a stone wall"—and at the same time slowing down the monetization of its reserves. How much the fear of being the source of a major oil accident/spill in the Bosporus weighed in the decision to pursue the land route is unknown, but company officials may have worried about parallels to the Exxon Valdez accident off Alaska in 1989; that incident greatly damaged the reputation of the company and resulted in severe financial penalties. For an excellent survey of the estimated costs of the various pipeline proposals to carry ACG production, see Ronald Soligo and Amy Myers Jaffe, "The economics of pipeline routes: The conundrum of oil exports from the Caspian Basin," Baker Institute Working Paper, April 1998, hereafter Baker Institute, available online at http://www.rice.edu/energy/publications/docs/UnlockingtheAssets_EconomicsPipelineRoutes.pdf.

61. See note 3 for AIOC membership percentages of ownership. Devon and ExxonMobil have elected not to be members of the BTC consortium. The members of BTC are BP (30.1%), SOCAR (25%), Unocal (8.9%), Statoil (8.71%), TPAO (6.53%), Total (5%), ENI/Agip (5%), Itochu (3.4%), Hess (2.36%), Inpex (2.5%),

and ConocoPhillips (2.5%). Total, ENI, and ConocoPhillips are in BTC but not in AIOC; they have interests in Kazakhstan and may wish to preserve, through BTC ownership, the ability in the future to ship north Caspian volumes via BTC.

62. Future undersea pipelines will probably have to await agreement by all five littoral states on the legal status of the Caspian Sea.

63. Anonymous, "Georgia—Source of high risks," *Caspian Investor*, Vol. 7, Iss. 4, April 2004: 4–6.

64. DOE/EIA, in a 2004 review of energy in Turkey, questioned whether the line would be constructed at all. See *Country Analysis Brief*, "Turkey," May 2004, 8, available online at http://www.eia.doe.gov/emeu/cabs/turkey.html. Email correspondence between the author and a company representative indicate that the decision to go forward was made in 2003 and that the line is under construction.

65. The line is in fact a combination of new segments and refurbished existing Russian segments.

66. Anonymous, "BG has Karachaganak first oil," *Energy Economist*, August 2003, 26; Anonymous, "Kazakhstan," *Petroleum Economist*, Vol. 70, Iss. 8, August 2003: 32.

67. Anonymous, "Kazakhstan to begin oil supply by the "Odessa-Brody" pipeline in second half of 2004," *Media Monitoring Report*, April 30, 2004, available online from ISI Emerging Markets database (subscription required) at http://site.securities.com/doc.html?pc=KZ&doc_id=55938346. See also Eugene Khartukov and Ellen Starostina, "Russia's new pipelines will debottleneck exports, production," *Oil & Gas Journal*, Vol. 101, Iss. 38, October 6, 2003: 62–73, available online at http://proquest.umi.com/pqdweb?index=9&did=430704581&SrchMode=1&sid=14&Fmt=4&VInst=PROD&VType=PQD&RQT=309&VName=PQD&TS=1129475286&clientId=3921.

68. Anonymous, "EBRD set to finance prolonging of Odessa-Brody pipeline," *Alexander's Gas & Oil Connections,* Vol. 9, Iss. 5, March 10, 2004, available online at http://www.gasandoil.com/goc/news/ntc41060.htm.

69. Anonymous, "Ukrainian, Georgian premiers agree on Odessa-Brody oil pipeline usage," *FBIS*, February 28, 2005 (available online at https:// www.fbis.gov/portal/server.pt/gateway/PTARGS_0_979_246_203_0_43/http%3B/apps.fbis.gov%3B7011/fbis.gov/search/Search?action=viewDocument&holding=5177838). See also Anonymous, "ITAR-TASS: Ukraine, Poland agree to extend Odessa-Brody pipeline," *FBIS*, March 5, 2005 (available online at https://www.fbis.gov/portal/server.pt/gateway/PTARGS_0_979_246_203_0_43/http%3B/apps.fbis.gov%3B7011/fbis.gov/search/Search?action=viewDocument&holding=5186792), and Anonymous, "BBC monitoring: Profile of Ukrainian fuel, Energy Minister Ivan Plachkov," *FBIS*, March 4, 2005 (available online at https://www.fbis.gov/portal/server.pt/gateway/PTARGS_0_979_246_203_0_43/http%3B/apps.fbis.gov%3B7011/fbis.gov/search/Search?action=viewDocument&holding=5187177).

70. Caspian oil differs in quality from Iranian oil; northern Iranian refineries have been reconfigured to run high-sulfur Caspian crudes.

71. Hooman Peimani. "Central Asian states increase energy swap deals with Iran." *Central Asia–Caucasus Analyst*, June 30, 2004, available online at http://www.cacianalyst.org/view_article.php?articleid=2493.

72. Hooman Peimani, "Iran launches two Caspian initiatives," *Central Asia–Caucasus Analyst*, June 15, 2005, available online at http://www.cacianalyst.org/view_article.php?articleid=3372.

73. See Anonymous, "BTC could check Iran's swap plans," *Petroleum Intelligence Weekly*, Vol. XLIII, Iss. 19, May 10, 2004, 2.

74. Tankers transiting the Bosporus Straits are limited in size to 150,000 deadweight tons (DWT). Growing exports from Novorossiysk will mean either additional tankers, alternative deliveries within the Black Sea, or alternative pipelines to avoid the Bosporus.

75. The ports of Omisalj, Croatia, and of Vlore, Albania, are deepwater ports capable of accommodating large tankers. Either of these lines would require reversal, and are also caught up in Ukrainian politics and the uncertainty over the direction of flows should Odessa–Brody flow northward. See Anonymous, "Northern route to remain open for Caspian Sea crude oil," *Oil & Gas Journal*, Vol. 102, Iss. 10, March 8, 2004: 64.

76. Julian Lee, "FSU pipeline advisory," *Centre for Global Energy Studies*, No. 5, 2004, 12 March 2004, available online at http://www.cges.co.uk/pdf-lib/SampleFSUadvisory2004.pdf, 2.

77. Anonymous, "Russia backs Bosporus bypass pipeline," *EIG—Energy Compass*, June 4, 2004, available online from ISI Emerging Markets database (subscription required) at http://site.securities.com/doc.html?pc=RU&doc_id=57365535. See also Anonymous, "Transneft found another route bypassing Bosporus," *CIS Oil & Gas Report*, June 2, 2004, available online from ISI Emerging Markets database (subscription required) at http://site.securities.com/doc.html?pc=UZ&doc_id=57231909, and Anonymous, "TNK-BP and Tatneft to bypass Bosporous," *CIS Oil & Gas Report*, May 26, 2004, available online from ISI Emerging Markets database (subscription required) at http://site.securities.com/doc.html?pc=RU&doc_id=56944265.

78. Anonymous, "Turkey to build two oil pipelines to scale down tanker traffic," *Alexander's Gas & Oil Connections*, Vol. 9, Iss. 10, May 18, 2004, available online at http://www.gasandoil.com/goc/news/nte42057.htm.

79. Anonymous, "A shortcut for Russian oil to Asia," *Energy Security*, prepared by the Institute for the Analysis of Global Security, March 31, 2004, available online at http://www.iags.org/n0331044.htm.

80. Anonymous, "Russia, Greece and Bulgaria to sign memorandum on building pipeline," *Alexander's Gas & Oil Connections*, Vol. 10, Iss. 3, February 10, 2005, available online at http://www.gasandoil.com/goc/news/ntr50672.htm. See also Anonymous, "UPDATE: Bulgaria, Greece, Russia sign deal on Balkan pipe," *Dow Jones Newswires*, April 12, 2005.

81. See McCafferty and Kretzschmar, *op. cit.*

82. See, for example, Anonymous, "Kazakhstan: Bridging the Caspian," *Stratfor*, April 19, 2005, available online at http://www.stratfor.biz/Print.neo?storId=247324.

83. Anonymous, "Kazakhstan to supply 30 m t of oil per year to Transcaucasian pipeline," *BBC Monitoring*, May 26, 2005, available online from ISI Emerging Markets database (subscription required) at http://site.securities.com/doc.html?pc=KZ&doc_id=77419085&query=&hlc=.

84. Ian MacWilliam, "Kazakh-China oil pipeline opens," *BBC News*, December 15, 2005, available online at http://news.bbc.co.uk/2/hi/asia-pacific/4530426.stm.

85. See Anonymous, "Kazakhstan, China may formalize oil pipeline plan in May," *Interfax*, April 26, 2004, accessed April 28, 2004, at http://www.interfax.

com/com?item=Kaz&pg=0&id=5716363&req=,available online on *FBIS* at https://www.fbis.gov/portal/server.pt/gateway/PTARGS_0_979_246_203_0_43/http%3B/apps.fbis.gov%3B7011/fbis.gov/search/Search?action=viewDocument&holding=4299075, and Anonymous, "Kazakhstan-China oil pipeline to cost $700 mln," *Interfax*, April 22, 2004, accessed April 28, 2004, at http://www.interfax.com/com?item=Kaz&pg=0&id=5715397&req=,available on *FBIS* at <https://www.fbis.gov/portal/server.pt/gateway/PTARGS_0_979_246_203_0_43/http%3B/apps.fbis.gov%3B7011/fbis.gov/search/Search?action=viewDocument&holding=4284983>.

86. We will examine the gas pipeline rehabilitation proposals in subsequent sections that take in-depth looks at the energy picture in each of the countries contemplating increasing natural gas exports.

87. See Anonymous, "Gazprom made a choice between Yamal and Central Asia," *CIS Oil & Gas Report*, April 26, 2004, available online from ISI Emerging Markets database (subscription required) at http://site.securities.com/doc.html?pc=KZ&doc_id=55736251, and Anonymous, "Central Asia—Center gas pipeline to be modernized," *Oil Review*, April 16, 2004, available online from ISI Emerging Markets database (subscription required) at http://site.securities.com/doc.html?pc=KZ&doc_id=55480952.

88. This discussion draws on Anonymous, "Refining and marketing," *The Petroleum Economist*, Vol. 72, Iss. 7, July 2005, 39; Anonymous, "Work on Nabucco pipeline gets underway," *Alexander's Gas & Oil Connections*, Vol. 10, Iss. 16, August 31, 2005, available online at http://www.gasandoil.com/goc/news/ntc53567.htm; and Anonymous, "Gas from Shah Deniz to flow to Europe," *Alexander's Gas & Oil Connections*, Vol. 10, Iss. 15, August 17, 2005, available online at http://www.gasandoil.com/goc/news/ntc53362.htm.

89. Terry Adams, "Caspian oil realities," *Briefing Paper New Series No. 23*, The Royal Institute of International Affairs, September 2001, 1, available online at http://www.riia.org/pdf/briefing_papers/BP 23.pdf.

90. See Anonymous, "PetroKazakhstan takes steps to cut crude transportation costs," *The Oil Daily*, June 26, 2003, where a company representative indicates that rail costs have been as high as $14–14.50 per barrel. Article available online from ISI Emerging Markets database (subscription required) at http://site.securities.com/doc.html?pc=KZ&doc_id=43858820.

91. Adams, *op. cit.*, 1.

92. Terry Adams, "Caspian hydrocarbons, the politicisation of regional pipelines, and the destabilisation of the Caucasus," *Caucasian Regional Studies*, Vol. 5, Nos. 1 & 2, 2000, available online at http://www.ciaonet.org/olj/crs/crs_2000/crs00_adt01.html, 4. See also PetroKazakhstan slide 16 of presentation entitled "Corporate Presentation, CIBC World Markets Institutional Investor Conference," Whistler, BC, Canada, February 19, 2004, available online at http://events.onlinebroadcasting.com/cibcworldmarkets/021904/ ppts/petrokazakhstan.pdf, and "PetroKazakhstan takes steps to cut crude transportation costs," *op. cit.*

93. These calculations depend on the precise nature of the government "take" applied to each development project, including the government's share of production, income tax collections, royalties, bonuses, and other government levies. The conclusions on whether it is worthwhile for the private companies to pursue developments are drawn from Salameh, *op. cit.*, November 2003, 16.

94. See Jeanne M. Perdue, "Technology credited for new records," *Drilling and Production Yearbook*, March 2003, for noting that in March 2002 a Chirag well set a record for that year of 18,000 b/d. This is available online at http://www.eandpnet.com/pdf/Miscellaneous.pdf.

95. Pietro Cavanna, "Kazakhstan Project," ENI slide presentation dated October 3, 2003, available online at http://www.eni.it/eniit/eni/servlet/view/eni/upload/documentazione/area_investor_relations/_24f42_0_xoidcmWopk/Kazakhstan ProjectsCAVANNA.pdf.

96. Irina Denisova and Ivan Grigoryev, "Pipeline's price for progress," *Caspian Investor*, Vol. 7, Iss. 5, May 2004: 18–20.

97. Pipeline tariffs to Baku and Novorossiysk provided by DOE/EIA in email correspondence with the author, January 2, 2004.

98. Anonymous, "CPC to increase tariffs 3.3% from July," *Interfax*, March 17, 2004, accessed March 31, 2004, at http://www.interfax.com/com?id=5704491&item=Kaz, available on *FBIS* at <https://www.fbis.gov/portal/server.pt/gateway/PTARGS_0_979_246_203_0_43/http %3B/apps.fbis.gov%3B7011/fbis.gov/search/Search?action=viewDocument&holding=4016372>.

99. Anonymous, "CPC shareholders give green light to pipeline expansion," *New Europe*, March 7, 2005, available online from ISI Emerging Markets database (subscription required) at <http://site.securities.com/doc.html?pc=KZ&doc_id=69968905&query=>.

100. Cavanna, *op. cit.*

101. See PetroKazakhstan slide 16 of the corporate presentation to CIBC, *op. cit.*, available online at http://events.onlinebroadcasting.com/cibcworldmarkets/021904/ppts/petrokazakhstan.pdf.

102. See Sam Fletcher, "World oil production to rise by 16 million b/d by 2010," *op. cit.*

103. For an excellent review of the history of the Caspian states on seabed delimitation, see R. Hrair Dekmejian and Hovann H. Simonian, *Troubled Waters: The Geopolitics of the Caspian Region* (London, UK, and New York, NY: I.B. Tauris & Co Ltd, 2003).

104. For a discussion of the role of international law in the Caspian delimitation controversy, see Cesare P. R. Romano, "The Caspian Sea dispute: The role of international law," January 2000, online CEPMLP journal, available online at http://www.dundee.ac.uk/cepmlp/journal/html/vol5/article5–10.html.

105. See Hossein Askari, "Caspian oil development: The sooner the better," *Business Economics*, April 2003, 43–49, for a hypothetical breakdown under various oil prices, production levels, and a 10 percent discount rate, of the discounted future earning s for the littoral countries. The effects of development delay on earnings from disagreement on seabed ownership are clear. Available online at http://proquest.umi.com/pqdweb?index=0&did=000000348722891&SrchMode=1&sid=3&Fmt=6&VInst=PROD&VType=PQD&RQT=309&VName=PQD&TS=1087843932&clientId=3921.

106. Despite this knowledge, research published in western sources has been remarkably sparse over the years, but is now becoming more common, particularly with the initiation of the Caspian Energy Programme. Charles Blandy authored one of the earliest pieces in 1997, entitled "The Caspian: A catastrophe in the making—The destruction of a unique ecosystem," Conflict Studies Research Centre (Sandhurst,

UK). A year later Douglas Blum published "The Russian trade-off: Environment and development in the Caspian Sea," *Journal of Environment & Development*, Vol. 7, Iss. 3, September 1998. These two together captured most of earlier work on Caspian environmental issues and the extent of the damage. Since that time, not only does DOE/EIA issue a yearly report entitled "Caspian Sea region: Environmental issues," but a number of other authors have also addressed the topic. Among these are articles by Levent Hekimoglu, "Caspian Oil and the environment: Curse or cure?" in Michael P. Croissant and Bülent Aras, eds., *Oil and Geopolitics in the Caspian Sea Region* (Westport, CT: Praeger Publishers, 1999); Siamak Namazi, "The Caspian's environmental woes," and Hormoz Goodarzy, "Organizational response to the Caspian's environmental needs," both in Hooshang Amirahmadi, ed., *The Caspian Region at a Crossroad* (New York, NY: St. Martin's Press, 2000); and Igor Zonn, "The Caspian Sea: Threats to its biological resources and environmental security," in Gennady Chufrin, ed., *The Security of the Caspian Sea Region* (Oxford, UK: Stockholm International Peace Research Institute and Oxford University Press, 2001), 69–82.

107. Anonymous, "Caspian Sea in danger of turning into environmental dead zone," *Alexander's Gas & Oil Connections*, Vol. 9, Iss. 14, July 21, 2004, available online at http://www.gasandoil.com/goc/news/ntc42931.htm.

108. Anonymous, "Russia is worried that the Caspian countries are unaware of environmental and bio-resource issues," *AzerPress News*, March 18, 2004, available online from ISI Emerging Markets database (subscription required) at http://site.securities.com/doc.html?pc=AZ&doc_id=54295719. See also Anonymous, "Huge oil find 'threatens Caspian.'" *BBC News*. December 4, 2002, available online at http://news.bbc.co.uk/2/hi/science/nature/2540321.stm.

109. Anonymous, "Kazakhstan: Two oil wells leaking into Caspian since December 2003," *FBIS*, April 13, 2004, available online at https://www.fbis.gov/portal/server.pt/gateway/PTARGS_0_979_215_203_0_43/http%3B/apps.fbis.gov%3B7011/fbis.gov/search/Search?action=viewDocument&hold-ing=4223221.

110. As Chevron experienced in 2002, when Kazakhstan imposed a $70-million fine for what it alleged was inappropriate storage of sulfur at the Tengiz field. Chevron argued it was storing the sulfur at world-class standards. The matter was eventually resolved with Chevron's payment of a $7-million fine.

111. See "Framework convention for the protection of the marine environment of the Caspian Sea," http://www.caspianenvironment.org/report_technical.htm#frameconv.

112. Telephone conversation on June 3, 2005 between author and Fritz Schlingemann, Director and Regional Representative, UNEP Regional Office for Europe, and email correspondence with Elena Kvitsinskaia, staff member of the UNEP Regional Office for Europe, December 13, 2005, and February 27–28, 2006.

113. See Framework Convention, *op. cit.*

114. See, for example, http://www.caspianenvironment.org/report_technical.htm#frameconv for the action plans of Iran, Kazakhstan, and Azerbaijan.

115. The Caspian accounts for approximately 90 percent of the world's sturgeon stocks. See also Zonn, *op. cit.*, 69–82.

116. Antoine Blua, "Caspian: Influx of killer jellyfish threatens fish stocks," *Eurasianet*, July 3, 2004, available online at http://www.eurasianet.org/departments/environment/articles/pp070304_pr.shtml.

117. Gulnara Ismailova, "Over-fishing of sturgeon could harm Azerbaijan's caviar export," *Central Asia–Caucasus Analyst*, June 2, 2004, available online at http://www.cacianalyst.org/view_article.php?articleid=2418.

118. Paul Rimple, "Azerbaijan: Taking stock of sturgeon," *Eurasianet*, February 26, 2006, available online at http://www.eurasianet.org/departments/insight/articles/eav022406.shtml.

119. Katy Mclaughlin, "For caviar lovers, it's time to gulp," *The Wall Street Journal*, April 21, 2004, D3.

120. The "Dutch Disease" derives from changes in the Dutch economy in the aftermath of the 1959 discovery of North Sea gas in Dutch waters. The Dutch currency—the guilder—appreciated as a result of strong export sales of gas. The results made Dutch manufacturers of non-oil goods noncompetitive on world markets, employment fell, and deindustrialization and movement to the nontradable goods sector began. The problems are normally compounded by the volatility in oil prices, and the unwillingness of governments to rein in grandiose spending plans and link these plans to actual receipts.

121. Paul Stevens, "Resource impact—Curse or blessing? A literature survey," *CEPMLP Internet Journal Draft Working Paper*, April 22, 2003, available online at http://www.dundee.ac.uk/cepmlp/journal/html/Vol14/Vol14_1.pdf:1 (35).

122. In a conversation with Professor Stevens on May 22, 2004, the author and he discussed this percentage. In his view, basically any percentage equal to or exceeding 20–25 percent of merchandise trade is clear evidence of a skewed economy; his choice of 30 percent is thus a generous estimate.

123. United Press International, "Azerbaijan struggles to prevent 'Dutch disease,'" *World Peace Herald*, August 5, 2005, available online at http://www.wpherald.com/storyview.php?StoryID=20050805-041114-3183r.

124. See SOFAZ, "SOFAZ revenue and expenditure statement for 2005," available online at http://www.oilfund.az/inside.php?nID=131, and National Bank of Kazakhstan, "International reserves and assets of the National Oil Fund of Republic [*sic*] of Kazakhstan," available online at http://213.211.77.60cont/publish406836_2664.pdf.

125. See David. L. Goldwyn, "Extracting transparency," *Georgetown Journal of International Affairs*, Vol. 5, Iss. 1, Winter/Spring 2004: 9, available online at http://proquest.umi.com/pqdweb?index=0&did=572738961&SrchMode=1&sid=1&Fmt=3&VInst=PROD&VType=PQD&RQT=309&VName=PQD&TS=1129556368&clientId=3921.

CHAPTER 3

1. Central Intelligence Agency, "Azerbaijan," *2005 World Factbook*, available online at http://www.cia.gov/cia/publications/factbook/geos/aj.html.

2. Wood Mackenzie, *Azerbaijan: Country Overview*, May 2003, 11, reports that in 1594 a well was dug by hand to a depth of 35 meters. Marco Polo reported in 1264 that he had seen people gather oil from seeps in Baku. See also Natalia Antelava and Dima bit-Suleiman, "Oil Is Not Enough?" *Transitions Online*, September 25, 2003, available online at http://www.tol.cz/look/TOL/article.tpl?IdLanguage=1&IdPublication=4&NrIssue=50&NrSection=2&NrArticle=10714

&search=search&SearchKeywords=oil + is + not + enough&SearchMode=on&SearchLevel=0. An excellent history of the country is contained in Charles van der Leeuw's *Azerbaijan: A Quest for Identity* (New York, NY: St. Martin's Press, 2000), and in Avedis Bedros Hadjian, "Azerbaijan's Energy Policy and its Implications for Russian Security," in *The Politics of Caspian Oil*, Bülent Gokay, ed. (Houndsmills, UK, and New York, NY: Palgrave, 2001).

3. Wood Mackenzie, *Azerbaijan: Country Overview*, May 2003, 11–14.

4. The discussion of Azerbaijan's Ilham Aliev draws on a number of sources, including: Anonymous, "Azerbaijan: Aliyev legacy headed for limbo?" *Stratfor*, August 4, 2003, available online at http://www.stratfor.info/Story.neo?storyId=220764; Anonymous, "Azerbaijan elections: All in the family," *Stratfor*, October 16, 2003, available online at http://www.stratfor.biz.ezproxy6.ndu.edu/Story.neo?storyId=223646; Anonymous, "Azerbaijan politics: President Pinocchio," *The Economist*, Vol. 369, Iss. 8346, October 18, 2003, 42, available online at http://proquest.umi.com.ezproxy6.ndu.edu/pqdweb?index=1&did=664855511&SrchMode=1&sid=1&Fmt=3&VInst=PROD&VType=PQD&RQT=309&VName=PQD&TS=1129399912&clientId=3921; Anonymous, "Azerbaijan says it is in "no hurry" to settle Nagorno-Karabakh conflict," *Eurasianet*, February 27, 2004, available online at http://www.eurasianet.org/departments/insight/articles/eav022704a.shtml; Anonymous, "Azerbaijan's succession: Kim Jong-Ilham," *The Economist*, Vol. 368, Iss. 8336, August 7, 2003, 32, available online at http://proquest.umi.com.ezproxy6.ndu.edu/pqdweb?index=0&did=382362111&SrchMode=1&sid=2&Fmt=3&VInst=PROD&VType=PQD&RQT=309&VName=PQD&TS=1129400033&clientId=3921; Anonymous, "How will regime change in Georgia affect Azerbaijan?" *RFE/RL*, December 1, 2003, available online at http://www.rferl.org/reports/azerbaijan-report/2003/12/0-011203.asp; Anonymous, "Ilham Aliev," *Biography Resource Center Online* (Gale Group 2003), available online at http://galenet.galegroup.com/servlet/BioRC?vrsn=149&OP=contains&locID=wash60683&srchtp=name&ca=1&c=1&AI=1440489&NA=Ilham&ste=12&tbst=prp&tab=1&n=10&docNum=K1650003774&bConts=41; Anonymous, "Nepotism in Central Asia," *The New York Times*, October 27, 2003, A22, available online at http://proquest.umi.com.ezproxy6.ndu.edu/pqdweb?index=0&did=431128381&SrchMode=1&sid=1&Fmt=3&VInst=PROD&VType=PQD&RQT=309&VName=PQD&TS=1129400631&clientId=3921; Anonymous, "Russian daily says all Azerbaijani millionaires 'tightly controlled' by Alievs," *FBIS*, October 23, 2003, available online (with U.S. government account) at https://www.fbis.gov/portal/server.pt/gateway/PTARGS_0_979_246_203_0_43/http%3B/apps.fbis.gov%3B7011/fbis.gov/search/Search?action=viewDocument&holding=2163601; Rose Brady, "Central Asia: All in the family," *Business Week*, Iss. 3855, October 27, 2003: 63, available online at http://proquest.umi.com.ezproxy6.ndu.edu/pqdweb?index=16&did=429370921&SrchMode=1&sid=1&Fmt=3&VInst=PROD&VType=PQD&RQT=309&VName=PQD&TS=1129400332&clientId=3921; Dan Brennan, "Like father, like son? Azerbaijan's bleak legacy," *Eurasianet*, December 23, 2003, available online at http://www.eurasianet.org/departments/insight/articles/eav122303.shtml; Susan B. Glasser, "President's son sweeps election in Azerbaijan," *The Washington Post*, October 16, 2003, A14, available online at http://proquest.umi.com.ezproxy6.ndu.edu/pqdweb?index

=4&did=424228391&SrchMode=1&sid=1&Fmt=3&VInst=PROD&VType=PQD&RQT=309&VName=PQD&TS=1129400816&clientId=3921; Susan B. Glasser, "Rise of leader's son sharpens Azerbaijan's identity crisis," *The Washington Post*, August 9, 2003, A8, available online at http://proquest.umi.com.ezproxy6.ndu.edu/pqdweb?index=31&did=382495151&SrchMode=1&sid=2&Fmt=3&VInst=PROD&VType=PQD&RQT=309&VName=PQD&TS=1129401008&clientId=3921; Katherine Hardin, "Elections in Azerbaijan: Another Aliev takes the helm," Cambridge Energy Resource Associates, Decision Brief, December 2, 2003 (Private Report); Kim Iskyan, "Playboy of the eastern world: Why you should care about Azerbaijan's presidential elections," *Slate*, October 13, 2003, available online at http://www.stratfor.info/Story.neo?storyId=220764; Marsha Lipman, "Birth of a dynasty?" *The Washington Post*, August 11, 2003, A17, available online at http://proquest.umi.com.ezproxy6.ndu.edu/pqdweb?index=11&did=382802321&SrchMode=1&sid=3&Fmt=3&VInst=PROD&VType=PQD&RQT=309&VName=PQD&TS=1129401110&clientId=3921; Stephen Mulvey, "Profile: Ilham Aliyev," *BBC News Online*, October 15, 2003, available online at http://news.bbc.co.uk/2/hi/europe/3194422.stm; Paula R. Newberg, "Oil-lubricated realpolitik deepens divisions in an ailing Caucasus," *The Los Angeles Times*, October 12, 2003, M2, available online at http://proquest.umi.com.ezproxy6.ndu.edu/pqdweb?index=0&did=422860631&SrchMode=1&sid=4&Fmt=3&VInst=PROD&VType=PQD&RQT=309&VName=PQD&TS=1129401264&clientId=3921; Claudia Rossett, "Potentate Jr.," *The Wall Street Journal*, November 6, 2002, A22, available online at http://proquest.umi.com.ezproxy6.ndu.edu/pqdweb?index=1&did=232704161&SrchMode=1&sid=5&Fmt=3&VInst=PROD&VType=PQD&RQT=309&VName=PQD&TS=1129401325&clientId=3921; and Igor Torbakov, "The geopolitical stakes are high as Azerbaijan holds presidential election," *Eurasianet*, October 15, 2003, available online at http://www.eurasianet.org/departments/insight/articles/eav101503.shtml.

5. Anonymous, "President Pinocchio," *op. cit.*, 42.

6. Anonymous, "Azerbaijan: Aliyev legacy headed for limbo?" *op. cit.*

7. Kim Iskyan, *op. cit.*

8. For a review of developments in the 1990s, see Cynthia Croissant, *Azerbaijan, Oil and Geopolitics* (Commack, NY: Nova Science Publishers, Inc., 1998).

9. See Anonymous, "SOCAR to develop structures dropped by foreign investors," *Interfax*, February 5, 2004, accessed June 6, 2004, at http://www.interfax.com/com?item=search&rpg=0&id=5685386&rea=lankoran, available on *FBIS* at https://www.fbis.gov/portal/server.pt/gateway/PTARGS_0_979_246_203_0_43/http%3B/apps.fbis.gov%3B7011/fbis.gov/search/Search?action=viewDocument&holding=3688448, and Anonymous, "JAOC pays Socar $4 million for pulling out of offshore project," *Prime-TASS News Wire*, May 12, 2004, available online from ISI Emerging Markets database (subscription required) at http://site.securities.com/doc.html?pc=RU&doc_id=56437306; email correspondence with a U.S. government expert, dated July 1, 2004.

10. Email correspondence with U.S. government expert dated June 24, 2004.

11. Caspian Events Limited and Azerbaijan Chamber of Commerce & Industry, "Caspian Oil & Gas 2004," 2003, 4; email correspondence with an industry

expert, September 29, 2003; email correspondence with a government expert, October 6, 2003; Wood Mackenzie, *Azerbaijan: Country Overview*, May 2003, State Department unclassified cables Baku 001410, 17 July 2003, and Baku 001487, 28 July 2003; and Noble Sky from Schlumberger website, and DOE/EIA, *Azerbaijan: Production-Sharing Agreements*, October 2004, available online at http://www.eia.doe.gov/emeu/cabs/azerproj.html. See also Anonymous, "SOCAR and Azenoil to sign contract on Binagadi bloc," *Alexander's Gas & Oil Connections*, Vol. 9, Iss. 21, October 28, 2004, available online at http://www.gasandoil.com/goc/company/cnc44326.htm, and Anonymous, "Azeri Parliament will ratify three PSAs in spring," *AzerPress News*, January 25, 2005, available online from ISI Emerging Markets database (subscription required) at http://site.securities.com/doc.html?pc=AZ&doc_id=67937393&query=azeri:parliament:will:ratify&hlc=ru.

12. Wood Mackenzie, *Azerbaijan: Country Overview*, May 2003, 34. It should be noted that DOE/EIA in 2004 estimated that Azeri onshore production for that year amounted to over 140,000 b/d, but reported to the author that 2005 daily onshore output was far lower at 35,000 b/d. See DOE/EIA, "Production-Sharing Agreements," October 2004, available online at http://www.eia.doe.gov/emeu/cabs/azerproj.html.

13. The remaining reserves are considered "technical" until additional sales contracts are signed. See Wood Mackenzie, *Azerbaijan: Country Overview*, May 2003, 28, 32.

14. See Tamam Bayatly, "BP Current Developments: First Pipe for BTC Pipeline Arrives in Baku," *Azerbaijan International*, Spring 2003, available online at http://www.azer.com/aiweb/categories/magazine/ai111_folder/111_articles/111_bp_development.html

15. The contract calls for an initial 2 bcm/y to Turkey, rising by 1 bcm/y each year until the peak of 6.6 bcm/y for 20 years. The Iranian contract builds from 5 bcm/y in 2003 to 10 bcm/y from 2007 to 2020; the renewal/expansion of the existing Russian contract calls for 14 bcm/y by pipeline through Bulgaria, while the trans-Black Sea "Blue Stream" Russian gas contract calls for an additional 16 bcm/y delivered to Samsun.

16. Anonymous, "Shah Deniz—biggest BP discovery since Prudhoe Bay," *Det Norske Veritas*, February 24, 2002, available online at http://www.dnv.com/publications/oilgas_news/articles/ShahDeniz.asp.

17. In the days of rosier gas demand forecasts, Turkey agreed to buy not only Shah Deniz gas (originally slated for 2004), but also to an expansion to 30 bcm/y of gas from Russia by 2009, to 10 bcm/y from Iran by 2007, and continuing purchases of over 5 bcm/y as LNG from Algeria and Nigeria together. EIA believes Turkey's gas contracts are twice the size of its forecast demand in 2005. See DOE/EIA, *Country Analysis Brief*, "Turkey," May 2004, available online at http://www.eia.doe.gov/emeu/cabs/turkey.html.

18. *Ibid.*

19. Anonymous, "Take-or-pay gas supplies may swamp Turkey's demand," *Oil & Gas Journal Online*, August 18, 2003, available online at http://ogj.pennnet.com/articles/article_display.cfm?Section=Archives&Article_Category=Gener&ARTICLE_ID=184659&KEYWORD=take%2Dor%2Dpay%20AND%20turkey.

20. The Greeks, the Azeris, and the Turks have considered plans for transporting up to 30 bcm/y of Azeri gas to Europe. Greece and Turkey signed a protocol of intent to build a 280-km line connecting their pipeline systems. See Anonymous, "Gas: BP postpones Caspian decision," *Energy Economist*, Iss. 251, September 2002, 26, available online at http://proquest.umi.com/pqdweb?index=4&did=000000380718771&SrchMode=3&sid=1&Fmt=3&VInst=PROD&VType=PQD&RQT=309&VName=PQD&TS=1087910583&clientId=3921. Gas moving west from Turkey could eventually reach Italy, and possibly Austria (through Bulgaria, Romania, and Hungary (see DOE/EIA, *Country Analysis Brief*, "Turkey," May 2004, available online at http://www.eia.doe.gov/emeu/cabs/turkey.html).

21. See Anonymous, "Shahdeniz gas field to meet 15% of Turkey's needs in gas," *New Europe*, July 18, 2005, available online from ISI Emerging Markets database (subscription required) at http://site.securities.com.ezproxy6.ndu.edu/doc.html?pc=TR&doc_id = 81922987&query=gas%3Apipeline%3AGreece&hlc=tr. See also Anonymous, "Greece-Italy Pipe," *World Gas Intelligence*, Vol. XVI, Iss. 26, June 29, 2005, 5, available online from ISI Emerging Markets database (subscription required) at http://site.securities.com.ezproxy6.ndu.edu/doc_pdf?pc=TR&doc_id=80552193&query=gas%3Apipeline%3AGreece&hlc=tr. Turkey is also courting Israel as a potential new gas purchaser.

22. Email correspondence with industry expert, June 23, 2004. The delivered price of Iranian gas to Turkey is estimated at just over $1 per MMBtu, vs. $3.25 per MMBtu of Russian gas via the Blue Stream pipeline. Iranian gas thus has the price advantage for onward shipments to West European countries. See "Take-or-pay gas supplies may swamp Turkey's demand," *op. cit.*

23. Anonymous, "European Border Price Estimates," *World Gas Intelligence*, Vol. XVII, Iss. 4, January 25, 2006, 6, available online from ISI Emerging Markets database (subscription required) at http://site.securities.com/doc_pdf?pc=ZZ&doc_id=96800538; email correspondence with U.S. government expert on Russian gas prices dated June 23, 2004.

24. See Hilary McCutcheon and Richard Osbon, "Risks temper Caspian rewards potential," *Oil & Gas Journal*, Vol. 99, Iss. 52, December 24, 2001: 22 (5).

25. BP Caspian Development and Export Bulletin, *Regional Review*, January 2002, Section 3, 11, available online at http://www.caspiandevelopmentandexport.com/Downloads/RR/Eng/Doc1/RR3.pdf.

26. Georgia already imports some natural gas from Azerbaijan.

27. The discussion of Saakashvili and Georgia draws on a variety of materials, including: Anonymous, "Russia and Georgia: Smiles all round—till the next time," *The Economist*, Vol. 365, Iss. 8294, October 12, 2002: 41, available online at http://proquest.umi.com.ezproxy6.ndu.edu/pqdweb?index=35&did=208853211&SrchMode=1&sid=6&Fmt=3&VInst=PROD&VType=PQD&RQT=309&VName=PQD&TS=1129401451&clientId=3921; Anonymous, "Experts discuss Georgian presidential frontrunner's positive, negative features," *FBIS*, December 29, 2003, available online (with U.S. government account) at https://www.fbis.gov/portal/server.pt/gateway/PTARGS_0_979_215_203_0_43/http%3B/apps.fbis.gov%3B7011/fbis.gov/search/Search?action=viewDocument&holding=3141382;

Anonymous, "Georgia: Website reviews Saakashvili background, views on policy," *FBIS*, November 27, 2003, available online (with U.S. government account) at https://www.fbis.gov/portal/server.pt/gateway/PTARGS_0_979_215_203_0_43/http%
3B/apps.fbis.gov%3B7011/fbis.gov/search/Search?action=viewDocument&holding
=3153455; Anonymous, "Georgia's new president: The people expect," *The Economist*, Vol. 370, Iss. 8357, January 8, 2004: 30, available online at http://proquest.umi.com.ezproxy6.ndu.edu/pqdweb?index=333&did=522820861
&SrchMode=1&sid=8&Fmt=3&VInst=PROD&VType=PQD&RQT=309&V
Name=PQD&TS=1129401873&clientId=3921; Anonymous, "Mikhail Saakashvili," *Biography Resource Center Online* (Gale Group, 2004), available online at http://galenet.galegroup.com/servlet/BioRC?vrsn=149&OP=starts&locID=
wash60683&srchtp=name&ca=2&c=1&AI=1443597&NA=Saak&ste=12&
tbst=prp&tab=1&n=10&docNum=K1650004043&bConts=41; Anonymous, "Profile: Mikhail Saakashvili," *BBC News*, January 25, 2004, available online at <ttp://news.bbc.co.uk/1/hi/world/europe/3231852.stm; Anonymous, "Saakashvili opponents in Georgia say president is using anti-democratic methods to advance democracy," *Eurasianet*, March 1, 2004, available online at http://www.eurasianet.org/departments/insight/articles/eav030104_pr.shtml; Anonymous, "Year in review 2003: World-affairs Georgia," *Encyclopædia Britannica Online*, available online at http://search.eb.com/eb/article?eu=421922; Natalia Antelava and Giorgi Lomsadze, "Radical opposition party tops public opinion poll in Georgia as parliament election looms," *Eurasianet*, October 28, 2003, available online at http://www.eurasianet.org/departments/
rights/articles/eav102803.shtml; Irakly Areshidze, "Questions for Mikheil Saakashvili," *The Washington Post*, February 21, 2004, A19, available online at http://proquest.umi.com.ezproxy6.ndu.edu/pqdweb?index=1&did=548036551&
SrchMode=1&sid=43&Fmt=3&VInst=PROD&VType=PQD&RQT=309&
VName=PQD&TS=1108937214&clientId=3921; Peter Baker, "Called home to rebuild, Georgian led revolution; Reforms expected of U.S.-educated Saakashvili," *The Washington Post*, November 26, 2003, A1, available online at http://proquest.umi.
com.ezproxy6.ndu.edu/pqdweb?index=0&did=464763861&SrchMode=1&sid=
53&Fmt=3&VInst=PROD&VType=PQD&RQT=309&VName=PQD&TS=
1108937997&clientId=3921; Peter Baker and Susan B. Glasser, "Push for democracy has authoritarians unnerved," *The Washington Post*, January 26, 2004, A11, available online at http://proquest.umi.com.ezproxy6.ndu.edu/pqdweb?
index=0&did=530668241&SrchMode=1&sid=54&Fmt=3&VInst=PROD&
VType=PQD&RQT=309&VName=PQD&TS=1108938041&clientId=3921; Mark Berniker, "In Georgia's 'Rose Revolution,' red ink remains," *Eurasianet*, December 12, 2003, available online at http://www.eurasianet.org/departments/
business/articles/eav121203_pr.shtml; Sabine Freizer, "Georgia's constitutional amendments: A setback for democratization?" *Central Asia–Caucasus Analyst*, February 11, 2004, available online at http://www.cacianalyst.org/
view_article.php?articleid=2111; Charles King, "A rose among thorns: Georgia makes good," *Foreign Affairs*, Vol. 83, Iss. 2, March/April 2004: 13, available online at http://proquest.umi.com.ezproxy6.ndu.edu/pqdweb?index=0&did=
586448751&SrchMode=1&sid=52&Fmt=3&VInst=PROD&VType=PQD&
RQT=309&VName=PQD&TS=1108943747&clientId=3921; Charles King,

"Potemkin democracy: Four myths about post-Soviet Georgia," *The National Interest*, Iss. 66, Summer 2001: 93, available online at http://proquest.umi.com.ezproxy6.ndu.edu/pqdweb?index=2&did=73956675&SrchMode=1&sid=53&Fmt=4&VInst=PROD&VType=PQD&RQT=309&VName=PQD&TS=1108943804&clientId=3921; John Laughland, "Georgia on their mind," *The Guardian*, April 1, 2004, 22, available online at http://www.guardian.co.uk/comment/story/0,,1183335,00.html#article_continue; Alec Rasizade, "Georgia Meets the Test of its Independence," *Contemporary Review*, Vol. 284, Iss. 1658, March 2004: 139, available online at http://proquest.umi.com.ezproxy6.ndu.edu/pqdweb?index=0&did=595571751&SrchMode=1&sid=82&Fmt=3&VInst=PROD&VType=PQD&RQT=309&VName=PQD&TS=1108947324&clientId=3921; Paul Starobin, "Marching into the unknown: Georgia will be a test case of the new U.S. policy to shore up strategic but failing states," *Business Week*, Iss. 3804, October 21, 2002: 58, available online at http://www.businessweek.com/magazine/content/02_42/b3804065.htm; Nino Tetelashvili, "Georgia's revolution takes a dangerous turn," *Central Asia–Caucasus Analyst,* May 5, 2004, available online at http://www.cacianalyst.org/view_article.php?articleid=2336; and Daan van der Schriek, "Saakashvili's first 100 days: Bloom still on reforms, but critics voice concern," *Eurasianet*, May 10, 2004, available online at http://www.caspianbusinessnews.com/preview/energy/docs/jun30-2003.html.

28. Anonymous, "Georgia: Flirting with Tehran to entice Washington," *Stratfor*, May 3, 2005, available online at http://www.stratfor.biz/Print.neo?storyId=247891.

29. Millennium Challenge Corporation, "Georgia and Millennium Challenge Corporation: Building a partnership for poverty reduction through sustainable economic growth," September 12, 2005, available online at http://www.mca.gov/public_affairs/fact_sheets/Georgia_compact_fact_sheet.shtml.

30. Zurab Zhvania was found dead in his apartment in February 2005, allegedly from carbon monoxide poison from a faulty space heater. His family and others have never accepted the official explanation, and assert foul play.

31. John Laughland, *op. cit.*, 22, reports that Saakashvili has said "he did not see the point of having any opposition deputies in the national parliament."

32. While oil and gas exploration are underway both in the Black Sea and onshore, new finds and subsequent production are not expected to be major sources of revenue to the government. CanArgo, a Canadian company, and Anadarko and Frontera of the U.S., have exploration and production rights in Georgia.

33. Anonymous, "Georgia," *International Petroleum Encyclopedia 2002* (Tulsa, OK: Pennwell Corporation, 2002), 135.

34. Andreas Billmeier, Jonathan Dunn, and Bert van Selm, "In the pipeline: Georgia's oil and gas transit revenues," IMF Working Paper WP/04/209, 2004, p. 8, available online at http://www.imf.org/external/pubs/ft/wp/2004/wp04209.pdf.

35. Anonymous, "Georgia," International Petroleum Encyclopedia 2002, op. cit., 134.

36. Andreas Billmeier, Jonathan Dunn, and Bert van Selm, *op. cit.*

37. State budget revenue for 2004 was estimated at $969 million, according to The Economist Intelligence Unit's *Country Report on Georgia*, May 2005, 18, converted at 1.83 lari to the U.S. dollar (12).

38. See Anonymous, "Half of Georgian section of Baku-Tbilisi-Ceyhan to be built by year-end," *Interfax*, March 14, 2003, accessed December 16, 2003, available online at http://www.mail-archive.com/antinato@topica.com/msg08435.html.

39. Fariz Ismailzade, "Economic ties bind Azerbaijan and Georgia," *Eurasianet*, November 12, 2003, available online at http://www.eurasianet.org/departments/insight/articles/eav121103.shtml.

40. Anonymous, "Quick takes: The Republic of Georgia," *Oil & Gas Journal*, Vol. 101, Iss. 2, January 13, 2003, 8, available online at http://ogj.pennnet.com/Articles/Article_Display.cfm?Section=ARCHI&ARTICLE_ID=165514&VERSION_NUM=1&p=7. Under the terms of the pipeline agreement, Azerbaijan, Georgia, and Turkey assume responsibility for ensuring pipeline security and the costs of compensating the BTC member companies for any losses incurred as a result of future social and environmental laws. Individual firms are responsible for operational integrity of the pipeline (See Anonymous, "Quick takes: Concerns are rising," *Oil & Gas Journal*, Vol. 101, Iss. 4, January 27, 2003: 9, available online at http://ogj.pennnet.com/Articles/Article_Display.cfm?Section=ARCHI&ARTICLE_ID=166832&VERSION_NUM=1&p=7.)

41. Anonymous, "Pipeline participants to pay Georgia for security," *Interfax*, October 20, 2004, accessed October 23, 2004, at http://www.interfax.com/com?item=Geor&pg=0&id=5763563&req=, available online on *FBIS* at <https://www.fbis.gov/portal/server.pt/gateway/PTARGS_0_979_246_203_0_43/http%3B/apps.fbis.gov%3B7011/fbis.gov/search/Search?action=viewDocument&holding=4923472>.

42. BP, *op. cit.*, June 2005.

43. For an exceptionally comprehensive discussion of developments in Kazakhstan, see Martha Brill Olcott, *Kazakhstan: Unfulfilled Promise* (Washington, DC: Carnegie Endowment for International Peace, The Brookings Institution Press, 2002).

44. The discussion of Kazakhstan and Nazarbayev draws on a number of sources, including Ibragim Alibekov, "Kazakhgate fires up Kazakhstan's elections campaign," *Eurasianet*, May 20, 2004, available online at http://www.eurasianet.org/departments/insight/articles/eav101404a_pr.shtml; Anonymous, "Corruption case could alter Kazakhstan's energy-export calculus," *Eurasianet*, June 2, 2004, available online at http://www.eurasianet.org/departments/business/articles/eav060204_pr.shtml; Anonymous, "Kazakhstan: In the name of the father," *The Economist*, Vol. 369, Iss. 8344, October 4, 2003: 41, available online at http://proquest.umi.com.ezproxy6.ndu.edu/pqdweb?index=1&did=417665571&SrchMode=1&sid=19&Fmt=3&VInst=PROD&VType=PQD&RQT=309&VName=PQD&TS=1108934533&clientId=3921; Anonymous, "Kazakhstan's ambassador in Russia resigns; Entry into politics suspected," *FBIS*, November 6, 2003, available online (with U.S. government account) at https://www.fbis.gov/portal/server.pt/gateway/PTARGS_0_979_246_203_0_43/http%3B/apps.fbis.gov%3B7011/fbis.gov/search/Search?action=viewDocument&holding=2593234; Anonymous, "Nazarbayev, Nursultan," in Clifford Thompson, ed., *Current Biography Yearbook 2000* (New York, NY, and Dublin, Ireland: The H.W. Wilson Company, October 2000), 426; Anonymous, "Nursultan Nazarbayev,"

Wikipedia, available online at http://en.wikipedia.org/wiki/Nursultan_Nazarbayev; Anonymous, "Nursultan Nazarbayev, President of Kazakhstan," *Worldmark Encyclopedia of the Nations: World Leaders* (Gale Group, 2003), available online at http://galenet.galegroup.com/servlet/BioRC?vrsn=149&OP=contains&locID=wash60683&srchtp=name&ca=1&c=1&AI=163030&NA=Nursultan&ste=12&tbst=prp&tab=1&n=10&docNum=K2303600374&bConts=47 - b_Essay; Nikola Krastev, "U.S/Uzbekistan: U.S. figure linked to Kazakh leaders faces corruption hearing," *RFE/RL*, June 3, 2004, available online at http://www.rferl.org/features/features_Article.aspx?m=06&y=2004&id=A918C05B-D472-4310-9AA7-76A7782F7AEA; Aldar Kusainov, "Wheels set in motion for dynastic transition in Kazakhstan," *Eurasianet*, September 16, 2003, available online at http://www.eurasianet.org/departments/insight/articles/eav091603_pr.shtml; Steve Levine, "Caspian intrigue: Odd family drama in Kazakhstan dims democratic hopes—Petroleum-rich land's ruler smacks down son-in-law and sundry challengers—Oil companies feel heat, too," *The Wall Street Journal*, September 12, 2002, A1, available online at http://proquest.umi.com.ezproxy6.ndu.edu/pqdweb?index=0&did=172597571&SrchMode=1&sid=55&Fmt=3&VInst=PROD&VType=PQD&RQT=309&VName=PQD&TS=1108944059&clientId=3921; Peter Marsh, "Long-serving Kazakh leader to stand again," *Financial Times*, July 3, 2003, 10; and Marlena Telvick, "Kazakhstan: United States vs. James H. Giffen," *Eurasianet*, May 31, 2004, available online at http://kazhegeldin.addr.com/2004en/engl_03_06_04.html.

45. "Nazerbayev, Nursultan," *op. cit.*, 426.

46. Dilyara Teshebayeva, "Kazakhstan plans vast expansion of payments to elderly, students, and state-sector employees," *Eurasianet*, May 5, 2006.

47. See Steve LeVine, "Caspian intrigue," *op. cit.*, for reporting on how a headless dog and the dog's head were left at the offices of the Republic newspaper and outside the newspaper editor's apartment.

48. Wood Mackenzie, *Kazakhstan: Country Overview*, November 2002, 26. Uzen is an old field; its production is in decline.

49. Ministry of Energy and Mineral Resources of the Republic of Kazakhstan, *Brief Extract, State Program for the Development of the Kazakhstan Sector of the Caspian Sea* (Astana, Kazakhstan: Ministry of Energy and Mineral Resources, Republic of Kazakhstan, May 2003, 1 (30)).

50. Anonymous, "Oil giants doubt bright Kazakhstani plans," *op. cit.*

51. Wood Mackenzie, *Kazakhstan: Country Overview*, *op. cit.*, 2002, 32, and *Kazakhstan: Country Overview*, November 2003, 29–30. The numbers reported in the 2003 report are somewhat lower than those in the 2002 report.

52. Wood Mackenzie, *Kazakhstan: Country Overview, op. cit.*, 2002, 36, and Kazakhstan: Country Overview, op. cit., 2003, 33.

53. This wording presumably means the concession could be extended at its expiration.

54. See Anonymous, "Tengiz sees production virtually flat y/y at 13.5 mn tons in 2006," *Intellinews*, February 13, 2006, available online from ISI Emerging Markets database, at http://site.securities.com/doc_pdf?pc=KZ&doc_id=97195728&auto=1&query=intellinews&db=en_1y_d&hlc=ru&range=365&sort_by=Date. Tonnes translated into mmbd at the rate of 7.9 barrels per tonne for Tengiz crude.

55. Gaffney, Cline & Associates, "Study of the oil and gas sectors in the Caspian with special emphasis on Kazakhstan and Azerbaijan," April, 2003, 19 (Glasgow, UK: Trade Partners UK, April 2003) available online at http://www.agcc.co.uk/documents/MAY_2003_GAFFNEY_CLINE_REPORT.pdf

56. Despite the field's discovery some years before the end of the Soviet Union, Soviet technology was not sufficiently advanced to deal with its development.

57. Trade Partners UK, "Kazakhstan oil and gas report," August 2002, 10, available online at http://www.agcc.co.uk/documents/Kazakhstan_Oil_and_Gas_Report_August_2002.doc.

58. Gaffney, Cline & Associates, *op. cit.*, 19–20.

59. Gaffney, Cline & Associates, *op. cit.*, 21.

60. DOE/EIA, *Country Analysis Brief*, "Kazakhstan," November 2004, 3, available online at http://www.eia.doe.gov/emeu/cabs/kazak.html, and Chevron Texaco, "Kazakhstan Fact Sheet," *op. cit.*

61. For a thorough discussion of these issues, see Katherine Hardin and Nurzhan Kapparov, "Kazakhstan's gas dilemma," *CERA Decision Brief*, Cambridge Energy Resource Associates (Cambridge, MA: January 2004), Private Report.

62. BP, BG, Statoil, and LUKoil at one point were Kashagan participants, but have since sold their interests.

63. Heather Timmons, "Kazakhstan wants a piece of its oil field," *New York Times*, June 29, 2004, W1.

64. Gabriel Kahn, Zaure Kistanova and Geoffrey T. Smith, "Kazakh government to buy into huge oil project," *The Wall Street Journal*, January 31, 2005, A16. See also Anonymous, "Eni's Group to sell 8.33 Kashagan stake, Kazakh minister says," *Bloomberg.com*, January 28, 2005, available online at http://www.bloomberg.com/apps/news?pid=10001099&sid=ausCQyrnR8xo&refer=energy.

65. Anonymous, "Kazakhstan: Green light at last for Kashagan," *Petroleum Economist*, Vol. 71, Iss. 4, April 2004: 37 (2).

66. Isabel Gorst, "Kazakhstan: Solving the export puzzle," *Petroleum Economist*, Vol. 69, Iss. 11, November 2002: 26 (3).

67. Nick Griffin and Caroline Cook, "Kashagan: The big sturgeon," *Deutsche Bank*, October 10, 2003: 21 (10). Private Report.

68. These expansions are also key to continued high exports of Russian crude. The Russian pipeline system is at about capacity, and Transneft has considered alternative export routes, including Ventspils and Murmansk, to reach world markets. See also Anonymous, "Transneft to increase capacity of Baltic Pipeline System to 50 million tons," *CIS Oil & Gas Report*, February 20, 2004, available online from ISI Emerging Markets database (subscription required) at http://site.securities.com/doc.html?pc=RU&doc_id=53238493, and DOE/EIA, *Country Analysis Brief*, "Baltic Sea region," January 2004, available online at http://www.eia.doe.gov/emeu/cabs/baltics.html.

69. Eric Watkins, "Kazakhstan, Azerbaijan to transport Kazakh crude through BTC pipeline," *Oil & Gas Journal Online*, June 6, 2003, available online at <http://ogj.pennnet.com/articles/article_display.cfm?Section=ARCHI&C=Trasp&ARTICLE_ID=178376&KEYWORDS=kazakhstan%2C%20azerbaijan%20to%20transport> (Subscription required).

70. *Ibid.* This article reports that some 380,000 b/d would move from Aktau to Baku for onward shipping on BTC. Yet none of the Kazakh producers have committed either volumes or a timetable for BTC shipments.

71. Charles van der Leeuw, "Kazakhs unveil oil plan," *Caspian Business News*, June 30, 2003, available online at http://www.caspianbusinessnews.com/preview/energy/docs/jun30–2003.html. The date slipped a year from 2005 to 2006. See also Anonymous, "KazMunaiGaz considering increasing carrying capacity of Atyrau-Samara pipeline to 20–25 million tones of oil per annum," *Caspian World*, April 30, 2004, available online at http://news.caspianworld.com/en/go/1915085189/-2121680150.

72. Anonymous, "Kazakhstan beats Russia to China," *Petroleum Intelligence Weekly*, Vol. LXIII, No. 43, October 25, 2004: 7.

73. Charles van der Leeuw, *op. cit.*, June 30, 2003.

74. Andrew Jack, "Nazarbayev considers oil pipeline to Gulf via Iran," *Financial Times (Asia Edition)*, May 26, 2004, 2.

75. See Anonymous, "Kazakhstani policies seen causing problems for foreign oil sector investors," *FBIS*, April 19, 2004, available online at https://www.fbis.gov/portal/server.pt/gateway/PTARGS_0_979_246_203_0_43/http%3B/apps.fbis.gov%3B7011/fbis.gov/search/Search?action=viewDocument&holding=4327482.

76. Anonymous, "ExxonMobil: New Kazakh tax rules discourage offshore devt," *Dow Jones Newswires*, October 6, 2004.

77. Christopher Pala, "Investors recoil from oil terms in Kazakhstan," *The New York Times*, October 27, 2004, W1.

78. Steve LeVine, "Kazakhstan hangs tough in oil flap—Clash with Chevron Texaco and ExxonMobil reveals new approach to foreigners," *The Wall Street Journal*, November 22, 2002, A11, available online at http://proquest.umi.com.ezproxy6.ndu.edu/pqdweb?index=0&did=242626161&SrchMode=1&sid=56&Fmt=3&VInst=PROD&VType=PQD&RQT=309&VName=PQD&TS=1108944244&clientId=3921.

79. Anonymous, "Kazakh oil giant favors pipeline through Russia rather than a US-backed facility," *Taipei Times*, March 28, 2004, available online at http://www.taipeitimes.com/News/biz/archives/2004/03/28/2003108099/print.

80. Anonymous, "Kazmunaigaz confirms tax problems at ChevronTexaco proj," *Dow Jones Newswires*, October 5, 2004.

81. Anonymous, "Kazakhstan: Agip KCO to pay fine for delayed extraction at Kashagan," *Caspian Investor*, April 2004, 21. Earlier this fine was thought to be in the $150–800 million range; see Anonymous, "Kashagan," *Petroleum Economist*, Vol. 70, Iss. 12, December 2003: 38. See also Vittorio Alessio, who reports in "ENI CEO says in talks about Kashagan project delay fines," *Dow Jones Business News*, October 9, 2003, that the Kazakh government was asking for $800 million in fines.

82. Julia Nanay, "The North Caspian PSA: Kashagan's development," *Middle East Economic Survey*, Vol. XLVII, No. 32, August 9, 2004, available online at http://www.mees.com/postedarticles/oped/a47n32d01.htm.

83. Leia Parker, "CNPC in $4.18 billion deal to buy PetroKazakhstan," *The Wall Street Journal Online*, August 22, 2005.

84. See http://www.petrokazakhstan.com, "PetroKazakhstan, Peters & Co. Limited, 2004 North American Oil & Gas Conference, September 15, 2004, Toronto, Ontario, Canada," Slide 22, giving lifting costs for the second quarter

of 2004 of $2.43 (USD) per barrel, and "PetroKazakhstan, Corporate Presentation, CIBC World Markets Institutional Investor Conference, Whistler, BC, February 17, 2005," Slide 17, giving lifting costs for the third quarter of 2004 of $2.26 (USD) per barrel.

85. Anonymous, "PetroKazakhstan takes steps to cut crude transportation costs," *op. cit.* See also Stephen O'Sullivan, Pavel Kushnir, and Olga Danilenko, *PetroKazakhstan—A Chinese Takeaway?* United Financial Group and Deutsche Bank, Private Report, August 4, 2004, 1.

86. The international press has reported on the cases and fines. See Anonymous, "Kazakhstan opens case against managers linked to Canadian oil firm," *FBIS*, October 27, 2004, available online at https://www.fbis.gov/portal/server.pt/gateway/PTARGS_0_979_246_203_0_43/http%3B/apps.fbis.gov%3B7011/fbis.gov/search/Search?action=viewDocument&holding=4935448; Anonymous, "Kazakh Supreme Court fines Canadian oil company 3.6 m dollars," *BBC Monitoring*, March 18, 2004, available online from ISI Emerging Markets database (subscription required) at http://site.securities.com/doc.html?pc=KZ&doc_id=54317102; Anonymous, "Kazakhstani policies seen causing problems for foreign oil sector investors," *op. cit*; Anonymous, "Antimonopoly agency intends to bring second suit against 'PetroKazakhstan' JSC," *Media Monitoring Report*, April 14, 2004, available online from ISI Emerging Markets database (subscription required) at http://site.securities.com/doc.html?pc=KZ&doc_id=55334611.

87. Overcoming the absence of international certification for locally manufactured products, updating the management systems of local providers, and providing both technical and English-language training to Kazakh nationals are continuing issues for TCO and other international oil firms and their international contractors.

88. U.S. Department of State cable, Almaty 005736, December 5, 2003 (Unclassified).

89. See Ivan Grigoryev and Maria Yaklovleva, "Key E&P projects in Northern Caspian Sea ratchet back the speed of development: Shifting into lower gear," *Caspian Investor*, Vol. 7, Iss. 5, May 2004: 27–30. See also Gerard Anderson, "The Kazakh upstream tax regime: Comparative analysis." *The International Investor's Magazine*, October 2004: 71 (9), available online at http://www.ey.com/global/download.nsf/Kazakhstan_E/Invest_Kz_2004_3RE.pdf/$file/Invest_Kz_2004_3RE.pdf.

90. See Ministry of Energy and Mineral Resources of the Republic of Kazakhstan, Brief Extract, State Program for the Development of the Kazakhstan Sector of the Caspian Sea, *op. cit.*, 13, 14, 28.

91. See Baker & McKenzie, "CIS legislative service," May/June 2003, 10, available online at http://www.bakernet.com/NR/ rdonlyres/ersmo3ijtybsxo3u53laitkeiyjk6dvokxbs7zeuxubu4tbwjyuxkfjt4vndl74jzmuve2a7mnts6g/LegServMayJune2003Engf.pdf. What appear to be similar and gross overestimates for the other littoral states are also quoted in the *State Program*.

92. See Anonymous, "Oil giants doubt bright Kazakhstani plans," *op. cit.*

93. The new 2004 tax regime—which includes taxes on produced oil—and increasingly hostile local/oblast conditions governing investment, including local content and locally trained and hired employees, could well slow the ambitious hopes of the national government to increase future production.

94. U.S. State Department cable, Almaty 000038, January 7, 2004 (Unclassified), and Azhar Kadrzhanova, "Kazakhstan: Caspian oil development plan," U.S. Commercial Service, July 3, 2003, available online at http://www.bisnis.doc.gov/bisnis/bisdoc/0307KZCaspOilProg.htm.

95. Anonymous, "Kazakh president signs off on PSA law for offshore projects," *Times of Central Asia*, July 20, 2005, available online from ISI Emerging Markets database (subscription required) at http://site.securities.com.ezproxy6.ndu.edu/doc.html?pc=KZ&doc_id=82344412&query=offshore%3APSA&hlc=ru. See also U.S. State Department cable, Almaty 000038, January 7, 2004 (Unclassified), and Charles van der Leeuw, *op. cit.*, June 30, 2003.

96. Kadrzhanova, *op. cit.*

97. See Anonymous, "Key E&P projects in Northern Caspian Sea ratchet back the speed of development: Shifting into lower gear," *Caspian Investor*, May 2004, 27 (4).

98. Anonymous, "Joint Investments of Russia and Kazakhstan in Kurmangazy project will amount to $22–23 billion," *CIS Oil and Gas Report*, July 8, 2005, available online from ISI Emerging Markets database (subscription required) at http://site.securities.com.ezproxy6.ndu.edu/doc.html?pc=RU&print=1&doc_id=81000791&query=Kurmangazy&hlc=ru, and Anonymous, "Kazakhstan, Russia sign oil production sharing agreement," *New Europe*, July 18, 2005, available online from ISI Emerging Markets database (subscription required) at http://site.securities.com.ezproxy6.ndu.edu/doc.html?pc=KZ&doc_id=81922990&query=Kazakhstan%3ARussia%3Asign&hlc=ru.

99. John Webb, "Russia's offshore Caspian projects: New dynamics in the wake of demarcation awards," *CERA Decision Brief*, Cambridge Energy Resource Associates (Cambridge, MA: August 2003), 1 (10), 10. Private Report.

100. See U.S. State Department cable, Almaty 000038, January 7, 2004 (Unclassified); Anonymous, "Lukoil: No oil from 1st test well at Caspian Tub-Karagan" *Dow Jones Newswires*, June 15, 2005; Anonymous, "LUKOIL plans to drill well at Tyub-Karagan sector in spring 2005," *Kazakhstan Today*, December 3, 2004, available online from ISI Emerging Markets database (subscription required) at http://site.securities.com/doc.html?pc=KZ&doc_id=65547721&query=. This source places total oil in place at around 2.4 billion barrels, or roughly 85 percent, with the remaining 15 percent of the field's resources in the form of natural gas. The Kazakh Minister of Energy earlier stated that the Tyub-Karagan projects held 825 million barrels of oil, but there is no evidence to confirm this number (see Anonymous, "Realization of Tyub-Karagan project to provide US$3.5 billion budget revenues," *Kazinform News*, December 29, 2003, available online from ISI Emerging Markets database (subscription required) at http://site.securities.com/doc.html?pc=KZ&doc_id=51337944. See also LUKoil's press release, "LUKoil enters Kazakhstan offshore projects," January 9, 2004, available online at http://www.lukoil.com/press.asp?div_id=1&id=2188&year=2004.

101. In January 2004, Shell downgraded its booked reserves by some 20 percent; its shares suffered on the stock markets of the world. Most international energy companies are conservative in estimating their reserves, and try to avoid an error or overstatement similar to that of Shell. Shell, however, is not the only company to revise its reserves position.

102. Janice Cowan, "Clearing the air Kazakhstan style," *Euromoney*, Iss. 381, January 2001, 86. See also Sabrina Tavernise and Christopher Pala, "Energy-rich Kazakhstan is suffering growing pains," *The New York Times*, January 4, 2003, C1, available online at http://proquest.umi.com/pqdweb?index=0&did=274106201&SrchMode=1&sid=2&Fmt=3&VInst=PROD&VType=PQD&RQT=309&VName=PQD&TS=1129473809&clientId=3921.

103. For a good review of Turkmenistan's energy, see Nancy Lubin, "Turkmenistan's energy: A source of wealth or instability?" in Robert Ebel and Rajan Menon, eds., *Energy and Conflict in Central Asia and the Caucasus* (Lanham, MD: Rowman & Littlefield Publishers, Inc., 2000).

104. The discussion of Turkmenistan and Niyazov draws on a number of sources, including Amnesty International's, "Turkmenistan Report 2003," covering events from January–December 2002, available online at http://web.amnesty.org/report2003/tkm-summary-eng, and its "Turkmenistan Report 2004," covering events from January–December 2003, available online at http://web.amnesty.org/report2004/tkm-summary-eng; Anonymous, "Ashkhabad's agitprop," *Foreign Policy*, Iss. 138, Sep/Oct 2003: 15; available online at http://proquest.umi.com.ezproxy6.ndu.edu/pqdweb?index=0&did=49239611&SrchMode=1&sid=3&Fmt=3&VInst=PROD&VType=PQD&RQT=309&VName=PQD&TS=1108931992&clientId=3921; Anonymous, "Asia: Time to choose; Turkmenistan," *The Economist*, Vol. 368, Iss. 8331, July 5, 2003, available online at http://proquest.umi.com.ezproxy6.ndu.edu/pqdweb?index=0&did=354864651&SrchMode=1&sid=6&Fmt=3&VInst=PROD&VType=PQD&RQT=309&VName=PQD&TS=1108932387&clientId=3921; Anonymous, "Health and dictatorship: Effects of repression in Turkmenistan," *The Lancet*, Vol. 361, January 4, 2003, available online at http://pdf.thelancet.com/pdfdownload?uid=llan.361.9351.health_and_human_rights.23869.1&x=x.pdf; Anonymous, "Saparmurad Niyazov, President of Turkmenistan," *Worldmark Encyclopedia of the Nations: World Leaders* (Gale Group, 2003), available online at http://galenet.galegroup.com/servlet/BioRC?vrsn=149&OP=contains&locID=wash60683&srchtp=name&ca=1&c=1&AI=162908&NA=Niyazov&ste=12&tbst=prp&tab=1&n=10&docNum=K2303600375&bConts=47 - b_Essay; Anonymous, "Saparmurat Niyazov," *Wikipedia*, available online at http://en.wikipedia.org/wiki/Niyazov; Anonymous, "The race to godhood," *Report/Newsmagazine (National Edition)*, Vol. 29, Iss. 17, September 2, 2002, available online at web30.epnet.com/resultlist.asp?tb=1&_ug=sid+760C67C7%2D4FCF%2D4D9F%2DA599%2D34559CCF1DCC%40sessionmgr5+dbs+afh%2Cmth%2Cbsh+cp+1+68A6&_us=hd+False+hs+False+or+Date+fh+False+ss+SO+sm+ES+dstb+ES+ri+KAAACB1C00001856+3352&_uso=hd+False+tg%5B2+%2D+tg%5B1+%2D+tg%5B0+%2DTI+st%5B2+%2D+st%5B1+%2D+st%5B0+%2Drace++to++godhood+db%5B2+%2Dmth+db%5B1+%2Dbsh+db%5B0+%2Dafh+op%5B2+%2DAnd+op%5B1+%2DAnd+op%5B0+%2D+4749&uh=1&sci=S1&sbt=1&lfr=ES; Anonymous, "Turkmenistan: President says society not yet ready for democracy," *FBIS*, October 26, 2003, available online (with U.S. government account) at https://www.fbis.gov/portal/server.pt/gateway/PTARGS_0_979_246_203_0_43/http%3B/apps.fbis.gov%3B7011/fbis.gov/search/Search?action=viewDocument&holding=2185462; Steve LeVine, "Turkmenistan's allegations may roil Central Asia," *The Wall Street Journal*, December 26, 2002, A8, available online at http://proquest.umi.com.ezproxy6.ndu.

edu/pqdweb?index=0&did=271191471&SrchMode=1&sid=57&Fmt=3&VInst=PROD&VType=PQD&RQT=309&VName=PQD&TS=1108944313&clientId=3921, and Ann McElvoy, "Ashkhabad guy: Niyazov's nasty reign," *The New Republic*, Vol. 210, Iss. 13, March 28, 1994, 15, available online at http://proquest.umi.com.ezproxy6.ndu.edu/pqdweb?index=0&did=5153390&SrchMode=1&sid=63&Fmt=3&VInst=PROD&VType=PQD&RQT=309&VName=PQD&TS=1108944976&clientId=3921.

105. Anonymous, "Turkmen president says leader should not be over 70," *FBIS*, May 7, 2004, available online (with U.S. government account) at https://www.fbis.gov/portal/server.pt/gateway/PTARGS_0_979_246_203_0_43/http%3B/apps.fbis.gov%3B7011/fbis.gov/search/Search?action=viewDocument&holding=4130647.

106. David Wallechinsky, "Who is the world's worst dictator? A new ranking of the contemptible," *The Washington Post*, Parade Magazine, February 13, 2005, 5.

107. Anonymous, "The race to godhood," *op. cit.*

108. See Anonymous, "Turkmenistan: 2003 investment climate statement," July 22, 2003, available online at http://www.bisnis.doc.gov/bisnis/bisdoc/030722TXICS.htm.

109. Anonymous, "Turkmen delegation discusses Islam and the prospects for political change," *The Eisenhower Institute*, February 8, 2002, available online at http://www.eisenhowerinstitute.org/presscenter/release16-02.htm. Shell, ExxonMobil, Bridas of Argentina, and International Petroleum Corporation of Canada pulled out of Turkmenistan some time ago. For a really damning assessment of Turkmenistan leadership and its repressive behavior, see Nancy Lubin, "Turkmenistan's energy: A source of wealth or instability," in *Energy and Conflict in Central Asia and the Caucasus*, Robert Ebel and Rajan Menon, eds. (Oxford, UK, and Lanham, MD: Rowman & Littlefield Publishers, Inc., 2000), 107–122.

110. Anonymous, "Turkmenistan: How healthy is Niyazov?" *Eurasianet*, June 21, 2005, available online at http://www.eurasianet.org/departments/insight/articles.eav062105_pr.shtml.

111. Wood Mackenzie, *Turkmenistan: Country Overview*, May 2004, 29.

112. Turkmenistan and Uzbekistan routinely claim that the other is stealing gas resources rightfully belonging to it.

113. Anonymous, "Turkmenistan lists ambitious plans for gas, oil development," *Oil & Gas Journal*, Vol. 100, Iss. 42, October 14, 2002: 43 (4), available online at http://proquest.umi.com/pqdweb?index=0&did=217200231&SrchMode=1&sid=3&Fmt=4&VInst=PROD&VType=PQD&RQT=309&VName=PQD&TS=1129473893&clientId=3921.

114. For a thorough assessment of Turkmenistan's fumbling of this proposal, see Bhamy Shenoy and William James, "Caspian energy exports: Turkmenistan fumbling opportunities afforded by Trans-Caspian Pipeline," *Oil & Gas Journal*, Vol. 98, Iss. 22, May 28, 2001: 70 (4).

115. India is no longer part of this proposal, although may yet reconsider participating.

116. *Interfax Petroleum Report*, June 10–17, 1994, as cited in Robert E. Ebel, *Energy Choices in the Near Abroad: the Haves and Have-nots Face the Future* (Washington, DC: The Center for Strategic and International Studies, 1997), 123.

117. Anonymous, "Worldwide look at reserves and production," *Oil & Gas Journal*, Vol. 103, No. 47, December 19, 2005, 24.

118. Anonymous, "Turkmenistan: Bold new production strategy," *Petroleum Economist*, Vol. 70, Iss. 11, November 2003: 32 (2), available online at http://proquest.umi.com/pqdweb?index=0&did=474691821&SrchMode=1&sid=4&Fmt=3&VInst=PROD&VType=PQD&RQT=309&VName=PQD&TS=1129473957&clientId=3921.

119. *Ibid.*

120. *Ibid.*, 33.

121. *Ibid.*, 32.

122. Prospects worldwide compete for scarce private capital, and the placement of that capital will be determined by political risk as well as price and geologic considerations.

123. Sergei Blagov, "Delay, competition mar Russia's Turkmenistan venture," *Eurasianet*, December 19, 2003, available online at http://www.eurasianet.org/departments/business/articles/eav121903_pr.shtml.

124. Press reports differ as to whether peak volumes will occur in 2007 or 2009, and whether these peak levels amount to 60, 70, 80, or 100 bcm/y.

125. Since 2005, gas prices in Western Europe have risen to close to $240 per thousand cubic meters, or about $6.80 per MMBtu.

126. This clause is likely to cause some contention, if it permits Russia the chance to overvalue the barter components, thus in effect lowering the price it pays for the gas.

127. Sergei Blagov, "Russia outmaneuvers Ukraine for Turkmen gas—For now," *Eurasianet*, April 21, 2005, available online at http://www.eurasianet.org/departments/business/articles/eav042105_pr.shtml.

128. As reported in Anonymous, "Turkmenistan: Bold new production strategy," *op. cit.*

129. See "Turkmenistan Project Open Society Institute," *Turkmenistan Weekly News Brief*, Iss. 7, February 10–16, 2006.

130. "Turkmenistan: Bold new production strategy," *op. cit.*

131. Alexander Vershinin, "Turkmenistan and China to sign gas deal," *Alexander's Gas & Oil Connections*, Vol. 10, Iss. 23, December 8, 2005, available online at http://www.gasandoil.com/goc/news/ntc54902.htm.

132. As reported in Anonymous, "Turkmenistan: Bold new production strategy," *op. cit.*

133. See Anonymous, "Turkmenistan faces challenges in export transportation options," *Oil & Gas Journal*, Vol. 100, Iss. 44, October 28, 2002, 48, and The Economist Intelligence Unit, *Country Report on Turkmenistan*, October 2003, 24.

134. Judy Clark, "Gazprom, partners to rehabilitate CAC pipeline, study awarded," *Oil & Gas Journal Online*, Nov. 12, 2003, available online (subscription required) at http://ogj.pennnet.com/articles/web_article_display.cfm?Section=Archives&Article_Category=Trasp&ARTICLE_ID=192189&KEYWORD=turkmen%20gas%20ukraine&x=y. See also Anonymous, "Central Asia—Center Gas Pipeline to be modernized," *Oil Review*, April 16, 2004, available online from ISI Emerging Markets database (subscription required) at http://site.securities.com/doc.html?pc=KZ&doc_id=55480952.

135. Judy Clark, *op. cit.* See also The Economist Intelligence Unit, *Country Report on Turkmenistan*, July 2003, 26.

136. Both The Economist Intelligence Unit and the *Petroleum Economist* have pointed out that Turkmenistan may have overcommitted not only producible gas reserves but gas pipeline transportation capacity as well.

137. Fereidun Fesharaki and Hassaan Vahidy, "Pakistan's gas discoveries eliminate import need," *Oil & Gas Journal*, Vol. 101, Iss. 6, January 28, 2002: 24–30.

138. This is the number that BP gives in its *Statistical Review of World Energy* issued in June 2005; the BP figure in the *Review* of June 2003 for Iranian proved reserves was 89.7 billion barrels. At the end of November 2003, Iran raised its estimates of its oil reserves to nearly 131 billion barrels, and BP in its 2004 *Statistical Review* reported 133.3 billion barrels for the end of 2003. This revision is believed to be in large part attributable to the huge new Azadegan onshore field. See also Anonymous, "Iran ramps up estimates of its oil reserves," *Alexander's Gas & Oil Connections*, Vol. 8, Iss. 24, December 11, 2003, available online at http://www.gasandoil.com/goc/news/ntm35095.htm. There is doubt, nonetheless, about the size of Iranian oil reserves, for the country has been known to announce reserves increases without demonstrating supporting evidence. See, for example, Mansour Kashfi, "Lack of data casts doubt on Iranian reserves hike," *Oil & Gas Journal*, Vol. 102, Iss. 41, November 1, 2004: 20–21. The Azadegan field is believed to contain over 26 billion barrels in total (not proved), far more than is estimated for Iran's future Caspian prospects. See DOE/EIA, *Country Analysis Brief*, "Iran," November 2003, 3, available online at http://www.eia.doe.gov/emeu/cabs/iran.html.

139. Whether this proposed pipeline were to terminate at Kharg Island in the Persian Gulf or at Bandar Abbas opposite the Straits of Hormuz makes little difference, since shipments from both would be at risk in case of either political or military difficulties.

140. BP, June 2005, *op. cit.* See also DOE/EIA, "OPEC oil production," *Short Term Energy Outlook*, September 2005, available online at http://www.eia.doe.gov/emeu/steo/pub/3atab.html.

141. Andrew Neff, "Caspian nations pursuing oil exports at greatly varying paces," *Oil & Gas Journal*, Vol. 103, Iss. 22, June 13, 2005: 38. See also DOE/EIA, *Country Analysis Brief*, "Iran," March 2005, available online at http://www.eia.doe.gov/emeu/cabs/iran.html, and *OGJ*, December 19, 2005, *op. cit.*

142. Neff, *op. cit.*, 38.

143. This discussion of the new Iranian President draws on Anonymous, "Mahmoud Ahmadinejad," *Biography Resource Center Online* (Gale Group, 2005), available online at http://galenet.galegroup.com.ezproxy6.ndu.edu/servlet/BioRC?vrsn=149&OP=starts&locID=wash60683&srchtp=name&ca=2&c=1&AI=1467073&NA=Ahmadinejad&ste=12&tbst=prp&tab=1&n=10&docNum=K1650004946&bConts=41, and Anonymous, "Cuba, Syria, Belarus join IAEA Board, boosting Iran," *Dow Jones Newswires*, September 29, 2005, available online at http://www.nasdaq.com/aspxcontent/NewsStory.aspx?cpath=20050929\ACQDJON200509291023DOWJONESDJONLINE000695.htm&selected=9999&selecteddisplaysymbol=9999&StoryTargetFrame=_top&mkt=WORLD&chk=unchecked&lang=&link=&headlinereturnpage=http://www.internat.

144. David E. Sanger, "U.S. challenge in Iran," *The New York Times*, June 26, 2005, 10.

145. Bahman Aghai Diba, "The legal regime of the Caspian Sea: Any changes in the positions of Iran?" *Alexander's Gas & Oil Connections*, Vol. 9, Iss. 5, March 10, 2004, available online at http://www.gasandoil.com/goc/news/ntc41059.htm.

146. David Townsend and Duncan Rushworth, "The Caspian angle," *Petroleum Economist*, Vol. 68, Iss. 9, September 2001: 28. The *Oil & Gas Journal* gives a range of from 0.0 to 0.1 billion barrels of oil. See the *OGJ* for December 20, 1999, and December 24, 2001. BP gives no reserves figures for either gas or oil for the Iranian Caspian area.

147. Discussions with industry representatives.

148. Hooman Peimani, "Iran intent on keeping pace with Caspian Basin energy development," *Eurasianet*, August 19, 2003, available online at http://www.eurasianet.org/departments/business/articles/eav081903_pr.shtml.

149. For an interesting discussion of these issues, see Askari, *op. cit.*

150. Anonymous, "Iran to launch oil production in Caspian Sea." *Alexander's Gas & Oil Connections*. Vol. 10, Iss. 4, February 24, 2005, available online at http://www.gasandoil.com/goc/news/ntc50810.htm.

151. Peimani, *op. cit.*, August 19, 2003.

152. Peimani, *op. cit.*, June 15, 2005.

153. See, for example, M. R. Farzanegan, "Iranian options most economically viable for exporting Caspian oil," *Oil & Gas Journal*, Vol. 101, Iss. 11, March 17, 2003: 22; Soligo and Jaffe in Baker Institute, *op. cit.*, 1998; and Joe Barnes and Ronald Soligo, "Baku-Ceyhan pipeline: Bad economics, bad politics, bad idea," *Oil & Gas Journal*, Vol. 96, Iss. 43, October 26, 1998: 29.

154. James Gavin, "Tehran drives for transit status," *Petroleum Economist*, Vol. 70, Iss. 12, December 2003: 5–6, available online at http://proquest.umi.com/pqdweb?index=0&did=520427081&SrchMode=1&sid=1&Fmt=3&VInst=PROD&VType=PQD&RQT=309&VName=PQD&TS=1129474229&clientId=3921. See also Hooman Peimani, "Central Asian states increase energy swap deals with Iran," *Central Asia–Caucasus Analyst*, June 30, 2004, available online at http://www.cacianalyst.org/view_article.phpaticleid=2493.

155. An earlier deal between Iran and Kazakhstan for swapping 120,000 b/d, signed in the mid-1990s, fell apart over differences in prices and quality specifications, in that Iranian refineries were unable to handle Kazakhstan's heavy-sulfur oil. See also Anonymous, "BTC could check Iran's swap plans," *op. cit*

156. Peimani, *op. cit.*, June 15, 2005. See also Charles van der Leeuw, "Iran peeks beyond the fence," *Caspian Business News*, August 4, 2003, available online at http://www.caspianbusinessnews.com/preview/energy/docs/aug4–2003.html; DOE/EIA, *Country Analysis Brief*, "Iran," November 2003, *op. cit.*; Anonymous, "Iran keen to cooperate with Caspian states," *Alexander's Gas & Oil Connections*, Vol. 9, Iss. 10, May 18, 2004, available online at http://www.gasandoil.com/goc/news/ntm42062.htm; and Anonymous, "Caspian crude swap project suspended due to price differences," *Payvand's Iran News*, February 5, 2005, available online at http://www.payvand.com /news/05/feb/1045.html.

157. Anonymous, "Kazakh shipping company transports more oil in 2005," *Asia Pulse*, February 7, 2006, available online at http://au.biz.yahoo.com/060207/17/ivyu.html.

158. Anonymous, "Iranian oil swap arrangements rising," *UPI*, January 9, 2006.

159. Hooman Peimani, "Central Asian states increase energy swap deals with Iran," *Central-Asia Caucasus Analyst*, June 30, 2004, available online at http://www.cacianalyst.org/view_article.php?articleid=2493.

160. Anonynmous, "Iran keen to conclude Caspian crude swap deals with Russia," *Asia Pulse*, June 8, 2005.

161. Anonymous, "India, Iran ink gas pipeline deal," *FBIS*, June 13, 2005, available on *FBIS* (with U.S. government account) at https://www.fbis.gov/portal/server.pt/gateway/PTARGS_0_979_246_203_0_43/http%3B/apps.fbis.gov%3B7011/fbis.gov/search/Search?action=viewDocument&holding=5382866.

162. DOE/EIA, *Country Analysis Brief*, "Turkey," July 2005, available online at http://www.eia.doe.gov/emeu/cabs/turkey.html.

163. Bülent Aliriza, "Turkey's Caspian energy quandary," *Caspian Energy Update, Center for Strategic and International Studies*, August 13, 2002, 5, available online at http://csis.org/turkey/ceu020813.pdf.

164. Anonymous, "Armenia-Iran pipeline may be extended to Ukraine, EU," *Interfax*, March 22, 2004, accessed March 22, 2004, at http://www.interfax.com/com?item=Arm+pg=0&id=5705536&req=,available on *FBIS* (with U.S. government account) at https://www.fbis.gov/portal/server.pt/gateway/PTARGS_0_979_246_203_0_43/http%3B/apps.fbis.gov%3B7011/fbis.gov/search/Search?action=viewDocument&holding=4063319; Anonymous, "Armenia and Iran ink gas, gas pipeline agreement," *Interfax*, May 14, 2004, accessed May 19, 2004, at http://www.interfax.com/com?item=Arm&pg=0&id=5721897&req=, available on *FBIS* online at https://www.fbis.gov/portal/server.pt/gateway/PTARGS_0_979_246_203_0_43/http%3B/apps.fbis.gov%3B7011/fbis.gov/search/Search?action=viewDocument&holding=4585880; Sergei Blagov, "Russia and Armenia: United by geopolitics, divided by energy resources," *Eurasianet*, May 17, 2004, available online at http://www.eurasianet.org/departments/business/articles/eav051704_pr.shtml; and Anonymous, "Iran to export gas to northern neighbors," *Alexander's Gas & Oil Connections*, Vol. 9, Iss. 10, May 18, 2004, available online at http://www.gasandoil.com/goc/news/ntm42046.htm.

165. American companies are forbidden to do business with Iran, although their foreign subsidiaries often do. Foreign companies who do more than $40 million worth of business with Iran are subject to sanctions under ILSA. The political question may be how willing the United States is to enforce sanctions and run the risk of straining its relations with European and other countries. The United States waived sanctions on Total's participation in the South Pars offshore gas field in exchange for securing French government pressure on Iran to stop support of terrorism.

166. Sam Fletcher, "US pressure disrupts Iran's S. Azadegan development," *Oil & Gas Journal*, Vol. 102, Iss. 35, September 20, 2004: 23, available online at http://proquest.umi.com/pqdweb?index=1&did=704003541&SrchMode=1&sid=2&Fmt=3&VInst=PROD&VType=PQD&RQT=309&VName=PQD&TS=1129474424&clientId=3921.

167. For an explanation of the lack of attractiveness of buyback contracts to foreign investors, see Jahangir Amuzegar, "Iranian oil buybacks: a formula no one likes," *Oil & Gas Journal*, Vol. 99, Iss. 35, August 27, 2001: 20, available online at http://proquest.umi.com/pqdweb?index=3&did=78957717&SrchMode=1&sid

=3&Fmt=4&VInst=PROD&VType=PQD&RQT=309&VName=PQD&TS=1129474467&clientId=3921. See also Alexander Brexendorff and Christian Ule, "Changes bring new attention to Iranian buyback contracts," *Oil & Gas Journal*, Vol. 102, Iss. 41, November 1, 2004: 16, available online at http://proquest.umi.com/pqdweb?index=0&did=735324151&SrchMode=1&sid=4&Fmt=4&VInst=PROD&VType=PQD&RQT=309&VName=PQD&TS=1129474525&clientId=3921.

168. See U.S. Department of State, "Background note: Iran," August 2004, 7, available online at http://www.state.gov/r/pa/ei/bgn/5314.htm.

169. Council on Foreign Relations, *Iran: Time for a New Approach*, July 2004, available online at http://www.cfr.org/pdf/Iran_TF.pdf.

170. For an excellent roundup of the situation in early 2004, see Isabel Gorst, "Whips and honey cakes," and Isabel Gorst, "Moscow flexes its muscles," both in *Petroleum Economist*, Vol. 71, Iss. 1, January 2004, 26–29, available online, respectively, at http://proquest.umi.com/pqdweb?index=0&did=543436911&SrchMode=1&sid=5&Fmt=3&VInst=PROD&VType=PQD&RQT=309&VName=PQD&TS=1129474638&clientId=3921, and at http://proquest.umi.com/pqdweb?index=3&did=543436901&SrchMode=1&sid=6&Fmt=3&VInst=PROD&VType=PQD&RQT=309&VName=PQD&TS=1129474698&clientId=3921.

171. Michael McFaul, Nikolai Petrov, and Andrei Ryabov, *Between Dictatorship and Democracy: Russian Post-Communist Political Reform* (Washington, DC: Brookings Institution Press, 2004).

172. Information provided by DOE/EIA.

173. Anonymous, "Russia to increase oil, gas production by nearly 50 percent over next 20 years," *FBIS*, November 13, 2003, available online at https://www.fbis.gov/portal/server.pt/gateway/PTARGS_0_979_246_203_0_43/http%3B/apps.fbis.gov%3B7011/fbis.gov/search/Search?action=viewDocument&holding=2581003.

174. Khartukov and Starostina, *op. cit.*, Table 4, 72.

175. See DOE/EIA, "Caspian Sea region: Key oil and gas statistics," August 2003, available online at *http://www.eia.doe.gov/emeu/cabs/caspstats.html*.

176. See Webb, *op. cit.*, 4.

177. Ivan Grigoryev, "Conflict of Caspian Pipeline Consortium shareholders' interests is ruining their business: Pipe ruptured by squabbling," *Caspian Investor*, Vol. 7, Iss. 2, February 2004: 21.

178. Anonymous, "SOCAR, Transneft discuss Baku-Novorossiisk supplies," *Interfax*, April 7, 2004, accessed June 6, 2004, at http://www.interfax.com/com?item=search&pg=0&id=5710482&req=socar, available on *FBIS* at https://www.fbis.gov/portal/server.pt/gateway/PTARGS_0_979_246_203_0_43/http%3B/apps.fbis.gov%3B7011/fbis.gov/search/Search?action=viewDocument&holding=4190407.

179. Email correspondence with DOE/EIA, January 2, 2004.

180. *Ibid.*

181. Anonymous, "Russian, Azerbaijani pres' talks focus on pipelines," *Dow Jones Newswires*, February 6, 2004.

182. Information provided on May 17, 2004, by U.S. government representative in Moscow.

183. Grigoryev and Yaklovleva, *op. cit.*

184. Anonymous, "Russia, Kazakhstan to sign $23 bln Caspian oil deal," *IRBIS—Financial Markets Information*, July 6, 2005, available online from ISI Emerging Markets database (subscription required) at http://site.securities.com.ezproxy6.ndu.edu/doc.html?pc=KZ&doc_id=80674406&query=%3Abln&hlc=ru.

185. Grigoryev and Yaklovleva, *op. cit.*

186. *Ibid.*

187. Anonymous, "LUKoil in Caspian," *Caspian Investor*, Vol. 7, Iss. 6, June/July 2004: 16.

188. See *OGJ* editors, "Study criticizes corporate governance in Russia," *Oil & Gas Journal Online*, October 5, 2004, available online at http://ogj.pennnet.com/articles/web_article_display.cfm?Section=OnlineArticles&ARTICLE_CATEGORY=GenIn&ARTICLE_ID=213249&x=y.

189. Fitch numbers have been widely reported in the press—see Anonymous, "Russian capital continues to fly abroad," *Mosnews.com*, June 14, 2005, available online at http://www.mosnews.com/money/2005/06/14/capitalflight.shtml. See also Anonymous, "Capital flight quadruples over Kremlin's continuing raids on corporate giants," *RIAN—Events in Russia*, February 7, 2005, available online from ISI Emerging Markets database (subscription required) at http://site.securities.com/doc.html?pc=RU&doc_id=68697400. See also Frank Brown, "Fear factor: In Vladimir Putin's Russia, worries about what comes next are stifling business," *Newsweek (International Edition)*, October 25, 2004, 28, available online at http://proquest.umi.com/pqdweb?index=1&did=717680491&SrchMode=1&sid=15&Fmt=3&VInst=PROD&VType=PQD&RQT=309&VName=PQD&TS=1129475692&clientId=3921. See also Gregory L. White, "Khodorkovsky declared guilty and sentenced to 9 years in jail," *The Wall Street Journal Online*, May 31, 2005, available online at http://online.wsj.com/article/0,SB111752952237546926,00.html, who suggests that $12 billion fled the country in 2004.

190. Gregory L. White, "Moscow restricts West's oil titans," *The Wall Street Journal*, February 11, 2005, A2, available online at http://proquest.umi.com/pqd-web?index=0&did=791892031&SrchMode=1&sid=16&Fmt=3&VInst=PROD&VType=PQD&RQT=309&VName=PQD&TS=1129475787&clientId=3921.

191. Guy Chazan, "Kremlin blocks Siemens purchase," *The Wall Street Journal*, April 14, 2005, A14, available online at http://proquest.umi.com/pqdweb?index=1&did=821506421&SrchMode=1&sid=17&Fmt=3&VInst=PROD&V-Type=PQD&RQT=309&VName=PQD&TS=1129475838&clientId=3921. It should be noted that had not ChevronTexaco bought Unocal, the U.S. Congress might have stepped in to block the sale of Unocal to CNOOC on similar grounds of "strategic interests."

192. Anonymous, "IMF is impressed with conditions of Russian economy," *The Russian Finance Report*, April 23, 2004, available online from ISI Emerging Markets database (subscription required) at http://site.securities.com/doc.html?pc=RU&doc_id=55682724.

193. See Anonymous, "Deputy minister sees Russia's Stability Fund to double by 2008," *Prime-Tass Business Newswire*, February 21, 2006, available

online from ISI Emerging Markets database (subscription required) at http://site.securit-ies.com.ezproxy6.ndu.edu/doc.html?pc=RU&doc_id=9790729&auto=1&query=Stability%3AFund&db=all_7d_d&hlc=en&range=3000&sort_by=Date.

194. Guy Chazan, "Russian taxman irks investors," *The Wall Street Journal*, November 10, 2004, A15, available online at http://proquest.umi.com/pqdweb?index=0&did=733317861&SrchMode=1&sid=1&Fmt=3&VInst=PROD&VType=PQD&RQT=309&VName=PQD&TS=1134482898&clientId=3921.

195. Information provided May 17, 2004, by a U.S. government representative in Moscow.

196. Anonymous, "CPC shareholders give green light to pipeline expansion," *New Europe*, March 7, 2005, available online from ISI Emerging Markets database (subscription required) at http://site.securities.com.ezproxy6.ndu.edu/doc.html?pc=RU&print=1&doc_id=69968905&query=&hlc=.

197. See Pavel Bakoulev, "PSA: Still alive in Russia?" *Caspian Investor*, Vol. 7, Iss. 4, April 2004: 27; Anonymous, "Russia," *Petroleum Intelligence Weekly*, Vol. XLIV, Iss. 14, April 4, 2005: 9, available online from ISI Emerging Markets database (subscription required) at http://site.securities.com/doc_pdf?pc=ZZ&doc_id=71872188; and Anonymous, "Russian tax law threatens future of PSAs," *EIG—Energy Compass*, May 16, 2003, available online from ISI Emerging Markets database (subscription required) at http://site.securities.com/doc.html?pc=RU&doc_id=42498187. It appears, however, that some of Putin's advisors have concluded either that foreign equity should be prohibited altogether, or that PSAs are inappropriate mechanisms, engendering corruption and speculation. See Anonymous, "Russia seizes energy reins," *EIG—Energy Compass*, February 6, 2004, and Anonymous, "Russian minister lambasts oil production sharing agreements," *BBC Monitoring*, April 1, 2004, both available online from ISI Emerging Markets database (subscription required) at, respectively, http://site.securities.com/doc.html?pc=RU&doc_id=52720499 and http://site.securities.com/doc.html?pc=RU&doc_id=54832469.

198. Roland Nash, "Taxing time for oil," *Alexander's Gas & Oil Connections*, Vol. 9, Iss. 3, February 10, 2004, available online at http://www.gasandoil.com/goc/news/ntr40660.htm.

199. Nash, *op. cit.*

200. Anonymous, "Russia favours higher oil export tariffs and extraction taxes," *Alexander's Gas & Oil Connections*, Vol. 9, Iss. 10, May 18, 2004, available online at http://www.gasandoil.com/goc/news/ntr42037.htm.

201. See Eric Watkins, "Watching the world: Spies, taxes, and business," *The Oil & Gas Journal*, Vol. 103, Iss. 16, April 25, 2005: 44, available online at http://ogj.pennnet.com/Articles/Article_Display.cfm?Section=ARCHI&ARTICLE_ID=226073&VERSION_NUM=2&p=7. See also Anonymous, "Oil export duty hike approved," *Interfax*, November 11, 2004, accessed November 11, 2004, at http://www.interfax.com/com?item=Rus&pg=0&id=5769156&req=,available online in *FBIS* (with U.S. government account) at https://www.fbis.gov/portal/server.pt/gateway/PTARGS_0_979_246_203_0_43/http%3B/apps.fbis.gov%3B7011/fbis.gov/search/Search?action=viewDocument&holding=4966341. See also Nash, *op. cit.*, and Erin Arvedlund, "Russia weighs tax shift toward oil sector," *The New York Times*, April 9, 2004, W1, available online at http://proquest.umi.com/pqdweb?index=0&did=612989531&SrchMode=1&sid=18&Fmt=3&VInst=

PROD&VType=PQD&RQT=309&VName=PQD&TS=1129476449&clientId=3921; and Anonymous, "Russian oil export duty to amount to $87.9 per ton from Oct. 1," *Alexander's Gas & Oil Connections*, Vol. 9, Iss. 18, September 21, 2004, available online at http://www.gasandoil.com/goc/news/ntr43860.htm.

202. See Andrew E. Kramer, "Moscow moves to restore a cap on price of gasoline," *New York Times*, February 16, 2006, C5.

203. See Anonymous, "It's make-or-break for Kovytka," *Petroleum Economist*, Vol. 71, Iss. 3, March 2004: 4–5, for a review of the political and tax uncertainties; available online at http://proquest.umi.com/pqdweb?index=30&did=609360391&SrchMode=3&sid=1&Fmt=3&VInst=PROD&VType=PQD&RQT=309&VName=PQD&TS=1129476695&clientId=3921&aid=1.

204. Anonymous, "TNK-BP won't lose Kovykta gas field license," *Alexander's Gas & Oil Connections*, Vol. 10, Iss. 13, July 6, 2005, available online at http://www.gasandoil.com/goc/company/cnr52701.htm.

205. Anonymous, "DJ China keen to advance Kovytka gas pipeline talks—TNK-BP," *Prime-Tass*, April 23, 2004, available online from ISI Emerging Markets database (subscription required) at http://site.securities.com/doc.html?pc=RU&doc_id=55689999&. See also Anonymous, "Russia could revoke license for giant gas field held by TNK-BP: Report," *Agence France-Presse*, January 30, 2004, available online from ISI Emerging Markets database (subscription required) at http://site.securities.com/doc.html?pc=RU&doc_id=52415113&, and Anonymous, "Giant Russian gas pipeline to China, South Korea gets green light," *Agence France-Presse*, November 14, 2003, available online from ISI Emerging Markets database (subscription required) at http://site.securities.com/doc.html?pc=RU&doc_id=49674710&.

206. See Anonymous, "Sinopec to Finance Exploration of a Part of Sakhalin-e Project," *The Russian Oil and Gas Report*, October 14, 2005, available online at http://web.lexis-nexis.com/universe/document?_m=5b07e73ac25d1b0c-2bca2ec6fde65688&_docnum=3&wchp=dGLbVzb-zSkVb&_md5=2aafb241991-104101278a3a832294433.

207. Anonymous, "Sakhalin-3 project strategic for Russia, resource minister says," *The Oil Daily*, April 2, 2004; Anonymous, "Sakhalin-3 fate uncertain," *The Oil Daily*, February 23, 2004; Anonymous, "Russia snubs Exxon; Refuses license for Sakhalin-3 project," *The Oil Daily*, January 30, 2004; and Anonymous, "Exxon changes tack on Sakhalin-3; Could accept non-PSA terms," *The Oil Daily*, September 24, 2003, all available online from ISI Emerging Markets database (subscription required) at, respectively, http://site.securities.com/doc.html?pc=RU&doc_id=54898568; http://site.securities.com/doc.html?pc=RU&doc_id=53335542;http://site.securities.com/doc.html?pc=RU&doc_id=52422066; and http://site.securities.com/doc.html?pc=RU&doc_id=47590598. See also Anonymous, "ExxonMobil hopes to become operator of Sakhlan-3 project," *The Russian Oil and Gas Report*, October 12, 2005, available online at http://web.lexis-nexis.com/universe/document?_m=44c2b02f3ae507d36fef940dffef9745&_docnum=4&wchp=dGLbVzz-zSkVb&_md5=c61645a501aef96b21d5be2081e91410, and Anonymous, "Rosneft is going to develop fields of Sakhalin-3 project," *The Russian Oil and Gas Report*, September 16, 2005, available online at http://web.lexis-nexis.com/universe/document?_m=44c2b02f3ae507d36fef940dffef9745&_docnum=1&wchp=dGLbVzz-zSkVb&_md5=6eaa568a76f666df0f51cdf9ba338b5d.

208. Andrew Somers, "Building blocks for increased trade and investment," *American Chamber of Commerce in Russia*, February 2004, available online at http://www.amcham.ru/pversion.php?item=973778168872990&att=2000.

209. This export proposal received a lot of attention in the press, yet it needs to be underscored that VLCC or ULCC tankers (very large crude carriers or ultra large crude carriers) are too big to dock in U.S. ports; they would have to lighter either in the Caribbean or at the Louisiana Offshore Oil Port (LOOP), thus diminishing the cost advantage over more distant Middle East cargoes.

210. See, for example, Mark N. Katz, "Assessing the Saudi-Russian summit: Much ado about nothing?" *Central Asia–Caucasus Analyst*, November 5, 2003, available online at http://www.cacianalyst.org/view_article.php?articleid=1872.

211. Sergei Blagov, "Russian oil companies to raise oil output as OPEC plans cut," *OPECNA News Service*, February 12, 2004, available online from ISI Emerging Markets database (subscription required) at http://site.securities.com/doc.html?pc=RU&doc_id=52875817.

212. Should it do so, however, Russia risks shattering the picture it has painted of itself that in the future it will be a reliable supplier to the United States at the time when some of its new projects at Murmansk come online. Cooperation with OPEC will cause the United States to doubt the sincerity of the reliability argument.

213. Anonymous, "Experts doubt that real costs of the Taishet-Nakhodka pipeline construction may be $12 billion," *CIS Oil & Gas Reports*, May 17, 2004, available online from ISI Emerging Markets database (subscription required) at http://site.securities.com/doc.html?pc=GE&doc_id=56570917.

214. For a thoughtful review of Russian pipelines and the role of Russian energy output in the world market, see Edward. C. Chow, "Russian pipelines: Back to the future," *Georgetown Journal of International Affairs*, Vol. 5, Iss. 1, Winter/Spring 2004: 27–33, available online at http://proquest.umi.com/pqdweb?index=9&did=572739031&SrchMode=1&sid=2&Fmt=3&VInst=PROD&VType=PQD&RQT=309&VName=PQD&TS=1129554445&clientId=3921.

215. Anonymous, "Russia will not allow any private gas or oil pipelines to be built," *Alexander's Gas & Oil Connections*, Vol. 9, Iss. 9, May 6, 2004, available online at http://www.gasandoil.com/goc/news/ntr41828.htm.

216. Anonymous, "Transneft President: Private pipeline is unthinkable in Russia," *Rosbalt News*, March 29, 2003, available online at http://www.rosbaltnews.com/2003/03/29/61690.html. The idea of an export terminal at Murmansk to serve U.S. as well as European markets has been around for years. As an alternative source of oil to that from the Middle East, it is attractive, and far closer to western markets. The problem is as noted earlier: the size of tankers. Large ones are too big for U.S. ports and would have to be lightered in the Caribbean or use the LOOP facility. Before Yukos fell from grace, Transneft was allegedly seeking U.S. government purchasing guarantees, perhaps as a means of retarding private company interest in building the Murmansk line. It should be noted that the United States has been importing both Russian crude oil and products since the country's independence. DOE/EIA estimates the 2005 9-month average level of U.S. imports from Russia at 227,000 b/d of crude and 201,000 b/d of products. See DOE/EIA, *Monthly Energy Review*, November 2005, Table 3.3g, 54, available online at http://www.eia.doe.gov/emeu/mer/pdf/pages/sec3_14.pdf. Should the

Murmansk line come to fruition, the likely crude exports to the United States might be 600,000 b/d or more.

217. We doubt the Turks would completely close the Straits to oil tanker traffic, since it would lead to a huge outcry about Turkey's attitude toward honoring treaties. It would also be politically unwise to offend EU members who use oil traveling out of the Black Sea if Turkey continues to want to join the EU.

218. U.S. Department of Energy, Office of Policy, "U.S.-Russia Energy Working Group overview and activities," October 2005, 1–8, provided to the author by DOE. This document summarizes the activities from the beginning of the dialogue.

219. U. S. Department of State, "Joint statement by President George W. Bush and President Vladimir V. Putin on the new U.S.-Russian energy dialogue," May 24, 2002, available online at http://www.whitehouse.gov/news/releases/2002/05/print/20020524–8.html.

220. U.S. Department of State, "Joint statement by President George W. Bush and President Vladimir V. Putin on U.S.-Russian energy cooperation," February 24, 2005, available online at http://www.state.gov/p/eur/rls/prsrl/2005/42696.htm.

CHAPTER 4

1. For recent views on China's energy position, see Center for Strategic & International Studies, "Implications of China's Energy Search," a conference on March 23, 2005, available online at http://www.csis.org/energy/050323_ChinaEnergyAgenda.pdf. For an older review of China's energy search, see Dru C. Gladney, "China's interest in Central Asia: Energy and ethnic security," in Ebel and Rajan, eds., *op. cit.*

2. Data provided by DOE/EIA.

3. For an excellent review of China's energy picture, see Martin Clark, "No going back," *Petroleum Economist*, Vol. 70, Iss. 12, December 2003: 7–10, available online at http://proquest.umi.com/pqdweb?index=15&did=726835051&SrchMode=3&sid=1&Fmt=3&VInst=PROD&VType=PQD&RQT=309&VName=PQD&TS=1129477250&clientId=3921&aid=1, and Derek Bamber, "Venturing into a new world," *Petroleum Economist*, Vol. 70, Iss. 12, December 2003: 11–12, available online at http://proquest.umi.com/pqdweb?index=29&did=520427071&SrchMode=3&sid=1&Fmt=3&VInst=PROD&VType=PQD&RQT=309&VName=PQD&TS=1129477318&clientId=3921&aid=1. See also Chapters 3 and 4 on oil and gas in Bernard D. Cole, *'Oil for the Lamps of China'—Beijing's 21st Century Search for Energy* (Washington, DC: Institute for National Strategic Studies, 2003). For a recent view of the reasons for why demand may fall in 2005, see Adam Sieminski, Nicholas Aldridge, Paul Sankey, and David Hurd, *Oil and Gas Abacus: Promise of Spring*, Deutsche Bank, February 23, 2005, 1 (35), 16. Private Report.

4. Chinese companies have either acquired interests, or are pursuing them, in Venezuela, Ecuador, Peru, Canada, Tunisia, Algeria, Gabon, Sudan, Malaysia, Iraq (under the Saddam regime), Iran, UAE, India, Russia, Azerbaijan, Kazakhstan, and most recently in Uzbekistan. See also Anonymous, "China energy: Efforts to diversify oil supplies," *Economist Intelligence Unit Industry Wire*, February 16,

2004, available online from ISI Emerging Markets database (subscription required) at http://site.securities.com/doc.html?pc=CN&doc_id=57209555.

5. Anonymous, "China begins building emergency oil stockpiles," *EIG—Energy Compass*, October 8, 2004, available online from ISI Emerging Markets database (subscription required) at http://site.securities.com/doc.html?pc=CN&doc_id=63072525 and information provided by DOE/EIA.

6. Mengdi Gu, "China wants more pipelines for improved oil import security," *Oil & Gas Journal*, Vol. 103, Iss. 1, January 3, 2005: 59, available online at http://ogj.pennnet.com/Articles/Article_Display.cfm?Section=ARCHI&ARTICLE_ID=218427&VERSION_NUM=1&p=7.

7. Anonymous, "China: Surging energy demand changes oil scene," *OxResearch*, February 26, 2004, available online at http://www.oxan.com/db/item.asp?NewsItemID=98105.

8. *Ibid.* See also Anonymous, "Get ready for the age of Chinese oil," *Petroleum Intelligence Weekly*, Vol. XLIII, Iss. 45, November 8, 2004: 1, available online from ISI Emerging Markets database (subscription required) at http://site.securities.com/doc_pdf?pc=ZZ&doc_id=64223823; James F. Wong, "China's booming economy needs energy, pipelines, and technology," *Oil & Gas Journal*, Vol. 101, Iss. 21, May 26, 2003: 55, available online at http://proquest.umi.com/pqdweb?index=10&did=342839321&SrchMode=1&sid=2&Fmt=3&VInst=PROD&VType=PQD&RQT=309&VName=PQD&TS=1129477712&clientId=3921; and John R. Brodman, "Energy (in)security in the 21st century," *IAEE Newsletter*, First Quarter 2004, 18, available online (subscription required) at http://www.iaee.org/documents/04win.pdf, 14–18.

9. Andrew Browne, Matt Pottinger, and Peter Wonacott, "China's expansion may be easing," *The Wall Street Journal*, June 11, 2004, A2, available online at http://proquest.umi.com/pqdweb?index=0&did=649858861&SrchMode=1&sid=5&Fmt=3&VInst=PROD&VType=PQD&RQT=309&VName=PQD&TS=1129477826&clientId=3921.

10. Donald Coxe, "The China syndrome," *Maclean's*, Vol. 117, Iss. 20, May 17, 2004: 29, available online at http://proquest.umi.com/pqdweb?index=38&did=636748351&SrchMode=3&sid=1&Fmt=3&VInst=PROD&VType=PQD&RQT=309&VName=PQD&TS=1129477973&clientId=3921&aid=1.

11. Anonymous, "The Asian giant stumbles: Will China come tumbling down?" *Stratfor*, May 12, 2004, available online at http://www.stratfor.biz/Story.new?storyId=231819. This article argues that China is likely to have a hard landing. Donald Coxe, *op. cit.*, that while there is an "emerging consensus" on a Chinese collapse, the country may yet avoid it.

12. Anonymous, "China: Economy—Down to earth with a bump?" *The Economist Intelligence Unit*, November 1, 2004. See also Sieminski *et. al.*, February 23, 2005, *op. cit.*

13. Certainly a 5 percent GDP growth rate is a very positive one, and one that many EU members would envy. Morris Goldstein and Nicholas R. Lardy, "What kind of landing for the Chinese economy?" *Policy Brief PB04–7*, Institute for International Economics, November 2004, available online at http://www.iie.com/publications/pb/pb04–7.pdf.

14. Peter Zeihan, "China's long march to bankruptcy," *Stratfor*, March 23, 2005, available online at http://www.Stratfor.biz/Story.neo?storyId=246127.

15. Anonymous, "Oil: The theory of relativity and rapid price change," *Stratfor*, July 19, 2005, available online at http://www.stratfor.biz.ezproxy6.ndu.edu/Story.neo?storyId=251977.

16. Correspondence with Vanessa Rossi, Oxford Economic Forum, May 18, 2004 and February 28–March 3, 2006, on China's energy demand.

17. Anonymous, "Russia looks to China to open East Siberia," *Petroleum Intelligence Weekly*, Vol. XLIII, Iss. 45, November 8, 2004: 3, available online from ISI Emerging Markets database (subscription required) at http://site.securities.com/doc_pdf?pc=ZZ&doc_id=64223823.

18. Anonymous, "Russia to build branch of oil pipeline to China," *Alexander's Gas & Oil Connections*, Vol. 10, Iss. 8, April 20, 2005, available online at http://www.gasandoil.com/goc/news/ntr51629.htm.

19. Anonymous, "Putin raises China's hopes over pipeline," *EIG—Energy Compass*, November 5, 2004, available online from ISI Emerging Markets database (subscription required) at http://site.securities.com/doc.html?pc=CN&doc_id=64165022.

20. Anonymous, "Russia looks to China to open East Siberia," *op. cit.* See also Anonymous, "Russia seeks low-cost loans from Japan to build oil pipeline," *Alexander's Gas & Oil Connections*, Vol. 10, Iss. 3, February 10, 2005, available online at http://www.gasandoil.com/goc/news/ntr50678.htm.

21. Anonymous, "Tokyo threatens to withdraw from Siberian pipeline financing," *Alexander's Gas & Oil Connections*, Vol. 10, Iss. 10, May 26, 2005, available online at http://www.gasandoil.com/goc/news/ntr52156.htm.

22. Anonymous, "Nelson Resources completes acquisition of a 50% participatory interest in the North Buzachi oil field development," *Kazakhstan Today*, February 5, 2004, available online from ISI Emerging Markets database (subscription required) at http://site.securities.com/doc.html?pc=KZ&doc_id=52617849.

23. Anonymous, "CNPC-Aktobemunaigaz launches gas plant in Kazakhstan," *Alexander's Gas & Oil Connections*, Vol. 8, Iss. 19, October 2, 2003, available online at http://www.gasandoil.com/goc/company/cnc34061.htm.

24. Katherine Hardin and Scott Roberts, "The fourth wave: Renewed Asian interest in the Caspian," *CERA Decision Brief*, Cambridge Energy Resource Associates (Cambridge, MA: August, 2005), 1 (13). Private Report. See the authors' Table 2 for the information on the Kuatamlonmunai field.

25. Gaye Christoffersen, "China's intentions for Russian and Central Asian oil and gas," *NBR Analysis*, The National Bureau of Asian Research, Vol. 9, Iss. 2, March 1998, 25, available online at http://www.nbr.org/publications/analysis/pdf/vol9no2.pdf.

26. *Ibid.*, 26.

27. Anonymous, "CNPC shelves China-Kazakhstan oil pipeline," *Oil & Gas Journal*, Vol. 97, Iss. 35, August 30, 1999: 44, available online at http://proquest.umi.com/pqdweb?index=0&did=44405359&SrchMode=1&sid=7&Fmt=3&VInst=PROD&VType=PQD&RQT=309&VName=PQD&TS=1129478702&clientId=3921. See also David Hurd, "China oil pipeline plans," *Oil & Gas Abacus: Can OPEC Pull It Off Again?* Deutsche Bank, November 3, 2003, 5, Private Report.

28. Hurd, *op. cit.*, 5. Initial capacity estimates are variously reported at 200,000–400,000 b/d but are now converging to 200,000 b/d. The full cost of the pipeline from the west of Kazakhstan to China is estimated at some

$3–3.5 billion. See also Anonymous, "Joining of Russia to Kazakhstan-China oil pipeline project will depend on operating conditions," *Transneft Press Center*, November 8, 2004, available online at http://www.transneft.ru/press/Default.asp?LANG=EN&ATYPE=8&PG=2&ID=6573; Anonymous, "Kazakh-China pipeline forges ahead," *EIG—Energy Compass*, October 8, 2004, available online from ISI Emerging Markets database (subscription required) at http://site.securities.com/doc.html?pc=CN&doc_id=63072528; Anonymous, "Construction of Kazakhstan-China oil pipeline to begin in 2004," *FBIS*, October 10, 2003, available online (with U.S. government account) at https://www.fbis.gov/portal/server.pt/gateway/PTARGS_0_979_215_203_0_43/http%3B/apps.fbis.gov%3B7011/fbis.gov/search/Search?action=viewDocument&holding=2059831; and Anonymous, "China, Kazakhstan ink crude pipeline deal—Report," *Dow Jones Newswires*, May 17, 2004.

29. Anonymous, "Kazakhstan, China set to speed up oil pipeline construction," *FBIS*, April 3, 2004, available online (with U.S. government account) at https://www.fbis.gov/portal/server.pt/gateway/PTARGS_0_979_246_203_0_43/http%3B/apps.fbis.gov%3B7011/fbis.gov/search/Search?action=viewDocument&holding=4153353.

30. PetroKazakhstan is also engaging in swaps with Iran from the northern Caspian port of Atyrau in Kazakhstan.

31. See discussions on this line and the uncertainties still surrounding it in Hugh Pope, "Balancing act gives China a key oil role," *The Wall Street Journal*, November 21, 2003, A10, available online at http://proquest.umi.com/pqdweb?index=2&did=456707141&SrchMode=1&sid=1&Fmt=3&VInst=PROD&VType=PQD&RQT=309&VName=PQD&TS=1129482322&clientId=3921; Ilan Berman, "The great game's newest player speaks Chinese," *The Wall Street Journal Online*, November 14, 2003; and Hurd, "China oil pipeline plans," *op. cit.*, 5.

32. See Anonymous, "West-East Gas Transmission Project," 2003, available online at http://www.infopetro.com/OilGasInChina/oilprojects3.htm. For volumes over 20 bcm/y, more compressor stations would need to be added or a parallel line laid.

33. See David A. Pietz, "China's emerging gas industry," *Oil & Gas Journal*, Vol. 98, Iss. 3, June 5, 2000, available online at http://proquest.umi.com/pqdweb?index=7&did=54974115&SrchMode=3&sid=1&Fmt=4&VInst=PROD&VType=PQD&RQT=309&VName=PQD&TS=1129482661&clientId=3921&aid=1. Details on buyers are sketchy; note that a capacity of 12 bcm/y by 2005 or of 20 bcm/y by 2010 will not succeed in reaching the stated goals. See also Anonymous, "China's oil and gas output increases steadily," *SynoSynergy*, January 30, 2004, available online at http://www.infopetro.com/Dissertation/ViewDissertation.asp?id=1136.

34. In fact, this has already happened, as gas is too expensive to win users to switch from coal. See Shai Oster and Patrick Barta, "About face: China stumbles in attempt to cut use of coal and oil," *The Wall Street Journal*, March 3, 2006, A1.

35. Carola Hoyos and Richard McGregor, "PetroChina ends talks on pipeline venture with foreign companies," *Financial Times*, August 4, 2004, 8.

36. Anonymous, "China, Kazakhstan may build gas pipeline," *Interfax*, November 5, 2004, accessed November 8, 2004, at http://www.interfax.com/

com?item=Kaz&pg=0&id=5768254&req=, available online (with U.S. government account) at *FBIS* at https://www.fbis.gov/portal/server.pt/gateway/PTARGS_0_979_246_203_0_43/http%3B/apps.fbis.gov%3B7011/fbis.gov/search/Search?action=viewDocument&holding=4954041.

37. The discussion of Karimov and Uzbekistan draws on a variety of materials including: Anonymous, "Central Asia—Leadership: Tricksters and tyrants," Canada & the World Backgrounder, *Encyclopedia Britannica Online*, March 2003, available online at http://search.eb.com/magazine/article?query=tricksters+tyrants&id=1&smode=3; Anonymous, "Asia: Day of the bully," *The Economist*, Vol. 361, Iss. 8245, October 27, 2001: 41, available online at http://proquest.umi.com.ezproxy6.ndu.edu/pqdweb?index=0&did=87441257&SrchMode=1&sid=5&Fmt=3&VInst=PROD&VType=PQD&RQT=309&VName=PQD&TS=1108932306&clientId=3921; Anonymous, "Islam Abduganievich Karimov," *Biography Resource Center Online* (Gale Group, 2003), available online at http://galenet.galegroup.com/servlet/BioRC?vrsn=149&OP=starts&locID=wash60683&srchtp=name&ca=2&c=2&AI=162984&NA=Karimov&ste=12&tbst=prp&tab=1&n=10&docNum=K2303600330&bConts=47 - b_Essay; Anonymous, "Islam Karimov," *Wikipedia*, available online at http://en.wikipedia.org/wiki/Islam_Karimov; Anonymous, "Profile of President Islam Karimov," *Human Rights Watch*, March 7, 2002, available online at http://www.hrw.org/press/2002/03/karimovprof.htm; Anonymous, "Such a popular president," *The Economist*, Vol. 362, Iss. 8258, February 2, 2002: 37, available online at http://proquest.umi.com.ezproxy6.ndu.edu/pqdweb?index=51&did=104103203&SrchMode=1&sid=30&Fmt=3&VInst=PROD&VType=PQD&RQT=309&VName=PQD&TS=1108936034&clientId=3921; Anonymous, "Survey: What's the rush?" *The Economist*, Vol. 368, Iss. 8334, July 26, 2003: 6, available online at http://proquest.umi.com.ezproxy6.ndu.edu/pqdweb?index=454&did=377543531&SrchMode=1&sid=29&Fmt=3&VInst=PROD&VType=PQD&RQT=309&VName=PQD&TS=1108935907&clientId=3921; Anonymous, "Uzbek Government Forces Closure of Local Soros Foundation," *Press Release—Open Society Institute*, April 18, 2004, available online at http://www.soros.org/newsroom/news/uzbekistan_20040418; Amnesty International, "Uzbekistan Report 2003," covering events from January to December 2002, available online at http://web.amnesty.org/report2003/uzb-summary-eng, and its "Uzbekistan Report 2004," covering events from January to December 2003, available online at http://web.amnesty.org/report2004/uzb-summary-eng; Anonymous, "Year in Review 2003: Uzbekistan," *Encyclopedia Britannica Article*, available online at http://cscs.umich.edu/~crshalizi/sloth/2003-08-13a.html; Antoine Blua, "Uzbekistan: EBRD To decide on future relationship," *RFE/RL*, April 5, 2004, available online at http://www.rferl.org/featuresarticle/2004/4/F5232698-55B2-44D9-B80C-A7E8B791F89F.html; Steve Crawshaw, "Clare Short helps out a tyrant," *New Statesman*, Vol. 16. Iss. 745, February 24, 2003: 17, available online at http://proquest.umi.com.ezproxy6.ndu.edu/pqdweb?index=0&did=298376851&SrchMode=1&sid=1&Fmt=3&VInst=PROD&VType=PQD&RQT=309&VName=PQD&TS=1129405453&clientId=3921; Chalain Hayes, "Uzbekistan Primer," September 16, 2001, available online at http://ndunet.ndu.edu/lib/terrorism/uzbekistan.pdf; Gulnoza Saidazimova, "Uzbekistan: Islam Karimov

vs. the Clans," *Eurasianet*, April 23, 2005, available online at http://www.eurasianet.org/departments/insight/articles/pp042305.shtml; David Stern, "Uzbekistan offers rich pickings for leader's daughter," *Financial Times*, August 19, 2003, 8; Robert Templer, "Tashkent dispatch: Steppe back," *New Republic*, Vol. 229, Iss. 7/8, August 18, 2003: 11, available online at https://ndu-gw.ndu.edu/exchange/CrandallM/Inbox/%20LexisNexis(R)%20Email%20Request%20%20(1861:66070886).EML/1_multipart_xF8FF_2_tashkent_dispatch.doc/C58EA28C-18C0-4a97-9AF2-036E93DDAFB3/tashkent_dispatch.doc?attach=1; Reed Tucker, "The five-minute guide to...the world's remaining tyrants," *Esquire*, Vol. 140, Iss. 2, August 2003: 110, available online at http://proquest.umi.com.ezproxy6.ndu.edu/pqdweb?index=37&did=370867201&SrchMode=3&sid=1&Fmt=3&VInst=PROD&VType=PQD&RQT=309&VName=PQD&TS=1129407218&clientId=3921&aid=1, and Kenneth Weisbrode, *Central Eurasia: Prize or Quicksand?* Adelphi Paper 338 (London, UK, and New York, NY: Oxford University Press, 2001).

38. See Anonymous, "A power struggle brews in Uzbekistan," Eurasianet, January 5, 2005, available online at http://www.eurasianet.org/departments/insight/articles/eav010505a_pr.shtml.

39. See, for example, Fiona Hill, "Uzbek fury," *The Wall Street Journal Online*, May 20, 2005, available online at http://online.wsj.com/article/0,SB111653700319638435,00.html; Martha Brill Olcott, "In Uzbekistan, the revolution won't be pretty," *The Washington Post*, May 22, 2005, B2, available online at http://proquest.umi.com.ezproxy6.ndu.edu/pqdweb?index=0&did=842916621&SrchMode=1&sid=1&Fmt=3&VInst=PROD&VType=PQD&RQT=309&VName=PQD&TS=1129407407&clientId=3921; and Anonymous, "Uzbekistan: No good political options in sight," *Eurasianet*, May 18, 2005, available online at http://www.eurasisanet.org/departments/insight/articles/eav051805_pr.shtml.

40. Oil production for the year 2005 provided by DOE/EIA, and gas production for the year 2004 from BP, June 2005, *op. cit.*

41. *Ibid.*

42. Anonymous, "Gazprom made a choice between Yamal and Central Asia," *op. cit.*, April 26, 2004.

43. Uzbek gas is high in sulfur, like that of its neighbors, and will require processing to remove it.

44. Anonymous, "Gazprom made a choice between Yamal and Central Asia," *op. cit.*, April 26, 2004.

45. See Anonymous, "Soviet-era pipeline to get partial makeover," *Stratfor*, May 14, 2003, available online at http://www.stratfor.biz/Print.neo?storyId=217335.

46. *Ibid.*

47. Anonymous, "Russian company to develop key Uzbek gas deposits," *FBIS*, September 22, 2003, available online at https://www.fbis.gov/portal/server.pt/gateway/PTARGS_0_979_246_203_0_43/http%3B/apps.fbis.gov%3B7011/fbis.gov/search/Search?action=viewDocument&holding=1676621. For details see Derek Brower, "Seeking export outlets," *Petroleum Economist*, Vol. 70, Iss. 11, November 2003: 34. More recent news on the Kandym project is contained in LUKoil's press release, "Lukoil Signs a Production Sharing Agreement on the

Kandym-Khausak-Shady Project in Uzbekistan," June 16, 2004, available online at http://www.lukoil.com/press.asp?div_id=1&id=2232&year=2004.

48. Anonymous, "Uzbekistan hopes to be key gas exporter," *Interfax*, October 15, 2004, accessed October 23, 2004, at http://www.interfax.com/com?item=Uzb&a=on&id=5762680, available on *FBIS* at <https://www.fbis.gov/portal/server.pt/gateway/PTARGS_0_979_246_203_0_43/http%3B/apps.fbis.gov%3B7011/fbis.gov/search/Search?action=viewDocument&holding=4904214>.

49. Wood Mackenzie, *Uzbekistan: Country Overview*, May 2003, 26, suggests that 87 percent of the country's hydrocarbon reserves are gas.

50. United States, Executive Office of the President, The White House, *A National Security Strategy for a New Century*, May 1997, 18, available online at http://clinton2.nara.gov/WH/EOP/NSC/Strategy.

51. These views are best captured in Joe Barnes, "U.S. national interests: Getting beyond the hype," in Yelena Kalyuzhnova, Amy Myers Jaffe, Dov Lynch, and Robin C. Sickles, eds., *Energy in the Caspian Region: Present and Future* (Hampshire, UK, and New York, NY: Palgrave Publishers Ltd., 2002), 212–233; in Amy Myers Jaffe, "US policy towards the Caspian region: Can the wish-list be realized?" in Gennady Chufrin, ed., *The Security of the Caspian Sea Region* (Oxford, UK: Stockholm International Peace Research Institute and Oxford University Press, 2001), 136–150; National Intelligence Council Conference Report, "Central Asia and the South Caucasus: Reorientations, Internal Transitions, and Strategic Dynamics," October 2000, available online at http://www.odci.gov/nic/confreports_asiacaucasus.html, and in Barnes and Soligo, *op. cit.*, 1998.

52. The Commission on America's National Interests, *America's National Interests* (Cambridge, MA: Harvard University, John F. Kennedy School of Government, Belfer Center for Science and International Affairs, July 2000), available online at http://bcsia.ksg.harvard.edu/BCSIA_content/documents/AmerNatInter.PDF.

53. *Ibid.*, 5.

54. *Ibid.*, 6.

55. *Ibid.*, 7.

56. *Ibid.*, 8.

57. After 9/11, security assistance to the region grew, and the U.S. military engagement and presence expanded to include new military air bases near Bishkek in Kyrgyzstan and Karshi-Khanabad in southern Uzbekistan near the Afghan border, although no explicit security guarantees were made in exchange for bases and staging areas. U.S. interest in military assistance to Uzbekistan cooled with that country's eviction of the United States from the Karshi-Khanabad base in mid-2005. Additional assistance includes the Georgia Train and Equip Program (GTEP), begun in May of 2002, to help develop a rapid reaction force to control territory and defend against potential terrorist threats in the Pankisi Gorge and elsewhere; waiver of Section 907 of the Freedom Support Act of 1992, which now permits U.S. aid to reach Azerbaijan; and U.S. funds for the first new military base of Kazakhstan's Western Military Command in Atyrau. Under consideration is the expansion of bilateral military cooperation with Azerbaijan in terms of protecting its sector of the Caspian Sea, perhaps a renewal of the Train and Equip Program in Azerbaijan, and the possibility of a U.S. or NATO military base on Azeri territory. U.S. engagement in joint exercises has in the past included naval exercises with Azerbaijan, emergency

rescue operations with Uzbekistan, and antiterrorism exercises with Kazakhstan. In addition, officers from the region participate in exercises in the United States each year. These exercises, with no Uzbek participation, are expected to continue in the future.

58. See Jim Lobe, "Uzbek leader becoming embarrassment to West," *OneWorld.net,* December 8, 2003, at http://us.oneworld.net/external/?url=http%3A%2F%2Fstory.news.yahoo.com%2Fnews%3Ftmpl%3Dstory%26cid%3D655%26ncid%3D655%26e%3D1%26u%3D%2Foneworld%2F20031208%2Fwl_oneworld%2F4536746991070890844.

Bibliography

Adams, Terry. "Caspian hydrocarbons, the politicisation of regional pipelines, and the destabilisation of the Caucasus." *Caucasian Regional Studies*. Vol. 5, No. 1 & 2, 2000.

———. "Caspian Oil Realities." *Briefing Paper New Series No. 23*. London, UK: The Royal Institute of International Affairs, 2001.

Aghai Diba, Bahman. "The legal regime of the Caspian Sea: Any changes in the positions of Iran?" *Alexander's Gas & Oil Connections*, Vol. 9, Iss. 5, March 10, 2004.

Alessio, Vittorio. "ENI CEO says in talks about Kashagan Project delay fines." *Dow Jones Business News*, October 9, 2003.

Alibekov, Ibragim. "Kazakhgate fires up Kazakhstan's elections campaign." *Eurasianet*, May 20, 2004.

Aliriza, Bulent. "Turkey's Caspian energy quandary." *Caspian Energy Update*, August 13, 2002. Washington, DC: Center for Strategic and International Studies, 2002.

Amnesty International. "Turkmenistan report 2004."

———. "Turkmenistan report 2003."

———. "Uzbekistan report 2004."

———. "Uzbekistan report 2003."

Amuzegar, Jahangir. "Iranian oil buybacks: A formula no one likes." *Oil & Gas Journal*, Vol. 99, Iss. 35, August 27, 2001: 20 (3).

Anderson, Gerard. "The Kazakh upstream tax regime: Comparative analysis." *The International Investor's Magazine*, October 2004: 71 (9).

Anonymous. "A power struggle brews in Uzbekistan." *Eurasianet*, January 5, 2005.
———. "A shortcut for Russian oil to Asia." *Institute for the Analysis of Global Security*, March 31, 2004.
———. "Antimonopoly agency intends to bring second suit against 'Petro Kazakhstan' JSC." *Media Monitoring Report*, April 14, 2004.
———. "Armenia and Iran ink gas, gas pipeline agreement." *Interfax*, May 14, 2004.
———. "Armenia-Iran pipeline may be extended to Ukraine, EU." *Interfax*, March 22, 2004.
———. "Ashkhabad's Agitprop." *Foreign Policy*, Iss. 138, September/October 2003: 15.
———. "Asia: Day of the bully." *The Economist*, Vol. 361, Iss. 8245, October 27, 2001: 84.
———. "Asia: Time to choose; Turkmenistan." *The Economist*, Vol. 368, Iss. 8331, July 5, 2003: 59.
———. "Azerbaijan: Aliyev legacy headed for limbo?" *Stratfor*, August 4, 2003.
———. "Azerbaijan elections: All in the family." *Stratfor*, October 16, 2003.
———. "Azerbaijan: President Pinocchio." *The Economist*, Vol. 369, Iss. 8346, October 18, 2003: 42.
———. "Azerbaijan says it is in 'no hurry' to settle Nagorno-Karabakh conflict." *Eurasianet*, February 27, 2004.
———. "Azerbaijan's succession: Kim Jong-Ilham." *The Economist*, Vol. 368, Iss. 8336, August 7, 2003: 32.
———. "Azeri parliament will ratify three PSAs in spring." *AzerPress News*, January 25, 2005.
———. "BBC monitoring: Profile of Ukrainian fuel, Energy Minister Ivan Plachkov." *FBIS*, March 4, 2005.
———. "BG has Karachaganak first oil." *Energy Economist*, Iss. 262, August 2003: 26.
———. "BTC could check Iran's swap plans." *Petroleum Intelligence Weekly*, Vol. XLIII, Iss. 19, May 10, 2004: 2.
———. "Capital flight quadruples over Kremlin's continuing raids on corporate giants." *RIAN—Events in Russia*, February 7, 2005.
———. "Caspian crude swap project suspended due to price differences." *Payvand's Iran News*, February 5, 2005.
———. "Caspian Sea in danger of turning into environmental dead zone." *Alexander's Oil & Gas Connections*, Vol. 9. Iss. 14, July 21, 2004.
———. "Central Asia—Center gas pipeline to be modernized." *Oil Review*, April 16, 2004.
———. "Central Asia—Leadership: Tricksters and tyrants." *Canada & the World Backgrounder*. Encyclopedia Britannica Online, March 2003.
———. "China begins building emergency oil stockpiles." *EIG—Energy Compass*, October 8, 2004.
———. "China: Economy—Down to earth with a bump?" *Economist Intelligence Unit*, EIU Business Asia, November 1, 2004.
———. "China energy: Efforts to diversify oil supplies." *Economist Intelligence Unit Industry Wire*, February 16, 2004.

———. "China, Kazakhstan ink crude pipeline deal—Report." *Dow Jones Newswires*, May 17, 2004.
———. "China, Kazakhstan may build gas pipeline." *Interfax*, November 5, 2004.
———. "China launches massive gas pipeline campaign." *Oil & Gas Journal*, Vol. 98, Iss. 23, June 5, 2000: 65.
———. "China: Surging energy demand changes oil scene." *OxResearch*. February 26, 2004.
———. "China's oil and gas output increases steadily." *SynoSynergy*. InfoPetro Technology Development, Ltd. January 30, 2004.
———. "CNPC-Aktobemunaigaz launches gas plant in Kazakhstan." *Alexander's Gas & Oil Connections*, Vol. 8, Iss. 19, October 2, 2003.
———. "CNPC shelves China-Kazakhstan oil pipeline." *Oil & Gas Journal*, Vol. 97, Iss. 35, August 30, 1999: 44.
———. "Construction of Kazakhstan-China oil pipeline to begin in 2004." *FBIS*, October 10, 2003.
———. "Corruption case could alter Kazakhstan's energy-export calculus." *Eurasianet*, June 2, 2004.
———. "CPC shareholders give green light to pipeline expansion." *New Europe*, March 7, 2005.
———. "CPC to increase tariffs 3.3% from July." *Interfax*, March 17, 2004.
———. "Cuba, Syria, Belarus join IAEA board, boosting Iran." *Dow Jones Newswires*, September 29, 2005.
———. "DJ China keen to advance Kovytka Gas Pipeline talks—TNK-BP." *Prime-Tass*, April 23, 2004.
———. "Deputy minister sees Russia's Stability Fund to double by 2008." *Prime-Tass Business Newswire*, February 21, 2006.
———. "EBRD set to finance prolonging of Odessa-Brody pipeline." *Alexander's Gas & Oil Connections*, Vol. 9, Iss. 5, March 10, 2004.
———. "Eni's group to sell 8.33 Kashagan stake, Kazakh minister says." *Bloomberg.com*, January 28, 2005.
———. "European border price estimates." *World Gas Intelligence*, Vol. XVII, Iss. 4, January 25, 2006: 6.
———. "Experts discuss Georgian presidential frontrunner's positive, negative features." *FBIS*, December 29, 2003.
———. "Experts doubt that real costs of the Taishet-Nakhodka pipeline construction may be $12 billion." *CIS Oil & Gas Reports*, May 17, 2004.
———. "Exxon changes tack on Sakhalin-3; Could accept non-PSA terms." *The Oil Daily*, September 24, 2003.
———. "ExxonMobil hopes to become operator of Sakhalin-3 project." *The Russian Oil and Gas Report*, October 12, 2005.
———. "ExxonMobil: New Kazakh tax rules discourage offshore devt." *Dow Jones Newswires*, October 6, 2004.
———. "Foreign PSAs help boost Turkmen oil production." *Oil & Gas Journal*, Vol. 100, Iss. 43, October 21, 2002: 50.
———. "Gas: BP postpones Caspian decision." *Energy Economist*, Iss. 251, September 2002: 26.

———. "Gas from Shah Deniz to flow to Europe." *Alexander's Gas & Oil Connections*, Vol. 10, Iss. 15, August 17, 2005.
———. "Gazprom made a choice between Yamal and Central Asia." *CIS Oil & Gas Report*, April 26, 2004.
———. "Georgia." *International Petroleum Encyclopedia 2002*. Tulsa, OK: Pennwell Corporation, 2002.
———. "Georgia: Flirting with Tehran to entice Washington." *Stratfor*, May 3, 2005.
———. "Georgia—Source of high risks." *Caspian Investor*, Vol. 7, Iss. 4, April 2004: 4 (3).
———. "Georgia: Website reviews Saakashvili background, views on policy." *FBIS*, November 27, 2003.
———. "Georgia's new president: The people expect." *The Economist*, Vol. 370, Iss. 8537, January 10, 2004: 30.
———. "Get ready for the age of Chinese oil." *Petroleum Intelligence Weekly*, Vol. XLIII, Iss. 45, November 8, 2004: 1, 4.
———. "Giant Russian gas pipeline to China, South Korea gets green light." *Agence France-Presse*, November 14, 2003.
———. "Greece-Italy pipe." *World Gas Intelligence*, Vol. XVI, Iss. 26, June 29, 2005: 5.
———. "Half of Georgian section of Baku-Tbilisi-Ceyhan to be built by year-end." *Interfax*, March 14, 2003.
———. "Health and dictatorship: Effects of repression in Turkmenistan." *The Lancet*, Vol. 361, January 4, 2003: 69 (2).
———. "How will regime change in Georgia affect Azerbaijan?" *RFE/RL*, December 1, 2003.
———. "Huge oil find 'threatens Caspian.'" *BBC News*, December 4, 2002.
———. "Ilham Aliev." *Biography Resource Center Online*. Washington, DC: Gale Group, 2003. Reproduced in *Biography Resource Center*. Farmington Hills, MI: Thomson Gale, 2004.
———. "IMF is impressed with conditions of Russian economy." *The Russian Finance Report*, April 23, 2004.
———. "India, Iran ink gas pipeline deal." *FBIS*, June 13, 2005.
———. "Iran keen to conclude Caspian crude swap deals with Russia." *Asia Pulse*, June 8, 2005.
———. "Iran keen to cooperate with Caspian states." *Alexander's Gas & Oil Connections*, Vol. 9, Iss. 10, May 18, 2004.
———. "Iran oil swap arrangements rising." *UPI*, January 9, 2006.
———. "Iran ramps up estimates of its oil reserves." *Alexander's Gas & Oil Connections*, Vol. 8, Iss. 24, December 11, 2003.
———. "Iran to export gas to northern neighbors." *Alexander's Gas & Oil Connections*, Vol. 9, Iss. 10, May 18, 2004.
———. "Iran to launch oil production in Caspian Sea." *Alexander's Gas & Oil Connections*, Vol. 10, Iss. 4, February 24, 2005.
———. "Islam Abduganievich Karimov." *Biography Resource Center Online*. Washington, DC: Gale Group, 2002. Reproduced in *Biography Resource Center*. Farmington Hills, MI: Thomson Gale, 2004.

———. "Islam Karimov." *Wikipedia*. Washington, DC, 2004.
———. "ITAR-TASS: Ukraine, Poland agree to extend Odessa-Brody Pipeline." *FBIS*, March 5, 2005.
———. "It's make-or-break for Kovytka." *Petroleum Economist*, Vol. 71, Iss. 3, March 2004.
———. "JAOC pays SOCAR $4 million for pulling out of offshore project." *Prime-Tass News Wire*, May 12, 2004.
———. "Joining of Russia to Kazakhstan-China oil pipeline project will depend on operating conditions." *Transneft Press Center*, November 8, 2004.
———. "Joint investments of Russia and Kazakhstan in Kurmangazy Project will amount to $22–23 billion." *CIS Oil & Gas Report*, July 8, 2005.
———. "Kashagan." *Petroleum Economist*, Vol. 70, Iss. 12, December 2003: 38.
———. "Kazakh-China pipeline forges ahead." *EIG—Energy Compass*, October 8, 2004.
———. "Kazakh oil giant favors pipeline through Russia rather than a US-backed facility." *Taipei Times*, March 28, 2004.
———. "Kazakh president signs off on PSA law for offshore projects." *Times of Central Asia*, July 20, 2005.
———. "Kazakh shipping company transports more oil in 2005." *Asia Pulse*, February 7, 2006.
———. "Kazakh Supreme Court fines Canadian oil company 3.6 m dollars." *BBC Monitoring*, March 18, 2004.
———. "Kazakhstan." *Petroleum Economist*, Vol. 70, Iss. 8, August 2003: 32 (2).
———. "Kazakhstan: Agip KCO to pay fine for delayed extraction at Kashagan." *Caspian Investor*, Vol. 7, Iss. 4, April 2004: 21.
———. "Kazakhstan beats Russia to China." *Petroleum Intelligence Weekly*, Vol. LXIII, No. 43, October 25, 2004: 7.
———. "Kazakhstan: Bridging the Caspian." *Stratfor*, April 19, 2005.
———. "Kazakhstan-China oil pipeline to cost $700 mln." *Interfax*, April 22, 2004.
———. "Kazakhstan, China may formalize oil pipeline plan in May." *Interfax*, April 26, 2004.
———. "Kazakhstan, China set to speed up oil pipeline construction." *FBIS*, April 3, 2004.
———. "Kazakhstan: Green light at last for Kashagan." *Petroleum Economist*, Vol. 71, Iss. 4, April 2004: 37 (2).
———. "Kazakhstan: In the name of the Father." *The Economist*, Vol. 369, Iss. 8344, October 4, 2003: 63.
———. "Kazakhstan opens case against managers linked to Canadian oil firm." *FBIS*, October 27, 2004.
———. "Kazakhstan, Russia sign oil production sharing agreement." *New Europe*, July 18, 2005.
———. "Kazakhstan: Solving the export puzzle." *Petroleum Economist*, Vol. 69. Iss. 11, November 2002: 26 (3).
———. "Kazakhstan to begin oil supply by the 'Odessa-Brody' pipeline in second half of 2004." *Media Monitoring Report*, April 30, 2004.
———. "Kazakhstan to supply 30 m t of oil per year to Transcaucasian pipeline." *BBC Monitoring*, May 26, 2005.

———. "Kazakhstan: Two oil wells leaking into Caspian since December 2003." *FBIS*, April 13, 2004.
———. "Kazakhstani policies seen causing problems for foreign oil sector investors." *FBIS*, April 19, 2004.
———. "Kazakhstan's ambassador in Russia resigns; Entry into politics suspected." *FBIS*, November 6, 2003.
———. "Kazmunaigaz confirms tax problems at ChevronTexaco proj." *Dow Jones Newswires*, October 5, 2004.
———. "KazMunaiGaz considering increasing carrying capacity of Atyrau-Samara pipeline to 20–25 million tonnes of oil per annum." *Caspian World*, April 30, 2004.
———. "Key E&P projects in northern Caspian Sea ratchet back the speed of development: Shifting into lower gear." *Caspian Investor*, May 2004: 27 (4).
———. "LUKoil enters Kazakhstan offshore projects." *LUKoil Press Release*, January 9, 2004.
———. "LUKoil in Caspian." *Caspian Investor*, Vol. 7, Iss. 6, June/July 2004: 16.
———. "LUKoil: No oil from 1st test well at Caspian Tyub-Karagan." *Dow Jones Newswires*, June 15, 2005.
———. "LUKoil plans to drill well at Tyub-Karagan sector in Spring 2005." *Kazakhstan Today*, December 3, 2004.
———. "LUKoil signs a production sharing agreement on the Kandym-Khausak-Shady Project in Uzbekistan." *LUKoil Press Release*, June 16, 2004.
———. "Mahmoud Ahmadinejad." *Biography Resource Center Online*. Washington, DC: Gale Group, 2005. Reproduced in *Biography Resource Center*. Farmington Hills, MI: Thomson Gale, 2005.
———. "Mikhail Saakashvili." *Biography Resource Center Online*. Washington, DC: Gale Group, 2004. Reproduced in *Biography Resource Center*. Farmington Hills, MI: Thomson Gale, 2004.
———. "Nazerbayev, Nursultan." In Clifford Thompson, ed., *Current Biography Yearbook 2000*. New York and Dublin, Ireland: The H. W. Wilson Company, October 2000.
———. "Nelson Resources completes acquisition of a 50% participatory interest in the North Buzachi oil field development." *Kazakhstan Today*, February 5, 2004.
———. "Nepotism in Central Asia." *The New York Times*, October 27, 2003: A22.
———. "Northern route to remain open for Caspian Sea crude oil." *Oil & Gas Journal*, Vol. 102, Iss. 10, March 8, 2004: 64 (2).
———. "Nursultan Nazarbayev." *Wikipedia*. Washington, DC, 2004.
———. "Nursultan Nazarbayev, President of Kazakhstan." *Worldmark Encyclopedia of the Nations: World Leaders*. Washington, DC: Gale Group, 2003. Reproduced in *Biography Resource Center*. Farmington Hills, MI: Thomson Gale, 2004.
———. "Oil export duty hike approved." *Interfax*, November 11, 2004.
———. "Oil giants doubt bright Kazakhstani plans." *RosBusinessConsulting—News Online*, October 9, 2003.
———. "Oil: The theory of relativity and rapid price change." *Stratfor*, July 19, 2005.

———. "PetroKazakhstan takes steps to cut crude transportation costs." *The Oil Daily*, June 26, 2003.
———. "Pipeline participants to pay Georgia for security." *Interfax*, October 20, 2004.
———. "Profile of President Islam Karimov." *Human Rights Watch*, March 7, 2002.
———. "Profile: Mikhail Saakashvili." *BBC News*, January 25, 2004.
———. "Putin raises China's hopes over pipeline." *EIG—Energy Compass*, November 5, 2004.
———. "Quick takes: Concerns are rising." *Oil & Gas Journal*, Vol. 101, Iss. 4, January 27, 2003: 9.
———. "Quick takes: The Republic of Georgia." *Oil & Gas Journal*, Vol. 101, Iss. 2, January 13, 2003: 8.
———. "Realization of Tyub-Karagan project to provide US $3.5 billion budget revenues." *Kazinform News*, December 29, 2003.
———. "Refining and marketing." *The Petroleum Economist*, Vol. 72, Iss. 7, July 2005: 39.
———. "Rosneft is going to develop fields of Sakhalin-3 project." *The Russian Oil and Gas Report*, September 16, 2005.
———. "Russia." *Petroleum Intelligence Weekly*, Vol. XLIV, Iss. 14, April 4, 2005: 9.
———. "Russia and Georgia: Smiles all round—Till the next time." *The Economist*, Vol. 365, Iss. 8294, October 12, 2003: 41.
———. "Russia backs Bosporus bypass pipeline." *EIG—Energy Compass*, June 4, 2004.
———. "Russia could revoke license for giant gas field held by TNK-BP: Report." *Agence France-Presse*, January 30, 2004.
———. "Russia favours higher oil export tariffs and extraction taxes." *Alexander's Gas & Oil Connections*, Vol. 9, Iss. 10, May 18, 2004.
———. "Russia fights Turkey over restricted Bosporus oil transports." *The Black Sea Pages*, May 5, 2003.
———. "Russia, Greece and Bulgaria to sign memorandum on building pipeline." *Alexander's Gas & Oil Connections*, Vol. 10, Iss. 3, February 10, 2005.
———. "Russia is worried that the Caspian countries are unaware of environmental and bio-resource issues." *AzerPress News*, March 18, 2004.
———. "Russia, Kazakhstan to sign $23 bln Caspian oil deal." *IRBIS—Financial Markets Information*, July 6, 2005.
———. "Russia looks to China to open East Siberia." *Petroleum Intelligence Weekly*, Vol. LXIII, Iss. 45, November 8, 2004: 3.
———. "Russia: Moody's upgrade reflects economic health." *OxResearch*, October 14, 2003.
———. "Russia seeks low-cost loans from Japan to build oil pipeline." *Alexander's Gas & Oil Connections*, Vol. 10, Iss. 3, February 10, 2005.
———. "Russia seizes energy reins." *EIG—Energy Compass*, February 6, 2004.
———. "Russia snubs Exxon; Refuses license for Sakhalin-3 project." *The Oil Daily*, January 30, 2004.
———. "Russia to build branch of oil pipeline to China." *Alexander's Gas & Oil Connections*, Vol. 10, Iss. 8, April 20, 2005.

———. "Russia to increase oil, gas production by nearly 50 percent over next 20 years." *FBIS*, November 13, 2003.
———. "Russia will not allow any private gas or oil pipelines to be built." *Alexander's Gas & Oil Connections*, Vol. 9, Iss. 9, May 6, 2004.
———. "Russian, Azerbaijani Pres' talks focus on pipelines." *Dow Jones Newswires*, February 6, 2004.
———. "Russian capital continues to fly abroad." *Mosnews.com*, June 14, 2005.
———. "Russian company to develop key Uzbek gas deposits." *FBIS*, September 22, 2003.
———. "Russian daily says all Azerbaijani millionaires 'tightly controlled' by Alievs." *FBIS*, October 23, 2003.
———. "Russian minister lambasts oil production sharing agreements." *BBC Monitoring*, April 1, 2004.
———. "Russian oil export duty to amount to $87.9 per ton from Oct. 1." *Alexander's Gas & Oil Connections*, Vol. 9, Iss. 18, September 21, 2004.
———. "Russian tax law threatens future of PSAs." *EIG—Energy Compass*, May 16, 2003.
———. "Saakashvili opponents in Georgia say president is using anti-democratic methods to advance democracy." *Eurasianet*, March 1, 2004.
———. "Sakhalin-3 fate uncertain." *The Oil Daily*, February 23, 2004.
———. "Sakhalin-3 project strategic for Russia, Resource Minister says." *The Oil Daily*, April 2, 2004.
———. "Saparmurad Niyazov, President of Turkmenistan." *Worldmark Encyclopedia of the Nations: World Leaders*. Washington, DC: Gale Group, 2003. Reproduced in *Biography Resource Center*. Farmington Hills, MI: Thomson Gale, 2004.
———. "Saparmurat Niyazov." *Wikipedia*. Washington, DC: 2003.
———. "Shah Deniz—Biggest BP discovery since Prudhoe Bay." *Det Norske Veritas*, February 24, 2002.
———. "Shahdeniz gas field to meet 15% of Turkey's needs in gas." *New Europe*, July 18, 2005.
———. "Sinopec to finance exploration of a part of Sakhalin-3 project." *The Russian Oil and Gas Report*, October 14, 2005.
———. "SOCAR and Azenoil to sign contract on Binagadi Bloc." *Alexander's Gas & Oil Connections*, Vol. 9, Iss. 21, October 28, 2004.
———. "SOCAR to develop structures dropped by foreign investors." *Interfax*, February 5, 2004.
———. "SOCAR, Transneft discuss Baku-Novorossiisk supplies." *Interfax*, April 7, 2004.
———. "Soviet-era pipeline to get partial makeover." *Stratfor*, May 14, 2003.
———. "Study criticizes corporate governance in Russia." *Oil & Gas Journal Online*, October 5, 2004.
———. "Such a popular president." *The Economist*, Vol. 362, Iss. 8258, February 2, 2002: 61.
———. "Survey: What's the rush?" *The Economist*, Vol. 368, Iss. 8334, July 26, 2003: 6.

———. "Take-or-pay gas supplies may swamp Turkey's demand." *Oil & Gas Journal Online*, August 18, 2003.

———. "Tengiz sees production virtually flat y/y at 13.5 mn tons in 2006." *Intellinews*, February 13, 2006.

———. "The Asian giant stumbles: Will China come tumbling down?" *Stratfor*, May 12, 2004.

———. "The race to Godhood." *Report/Newsmagazine (National Edition)*, Vol. 29, Iss.17, September 2, 2002.

———. "TNK-BP and Tatneft to bypass Bosporous." *CIS Oil & Gas Report*, May 26, 2004.

———. "TNK-BP won't lose Kovykta gas field license." *Alexander's Gas & Oil Connections*, Vol. 10, Iss. 13, July 6, 2005.

———. "Tokyo threatens to withdraw from Siberian pipeline financing." *Alexander's Gas & Oil Connections*, Vol. 10, Iss. 10, May 26, 2005.

———. "Transneft found another route bypassing Bosporus." *CIS Oil & Gas Report*, June 2, 2004.

———. "Transneft president: Private pipeline is unthinkable in Russia." *Rosbalt News*, March 11, 2003.

———. "Transneft to increase capacity of Baltic pipeline system to 50 million tons." *CIS Oil & Gas Report*, February 20, 2004.

———. "Turkey to build two oil pipelines to scale down tanker traffic." *Alexander's Gas & Oil Connections*, Vol. 9, Iss. 10, May 18, 2004.

———. "Turkmen delegation discusses Islam and the prospects for political change." *The Eisenhower Institute*, February 8, 2002.

———. "Turkmen president says leader should not be over 70." *FBIS*, May 7, 2004.

———. "Turkmenistan: 2003 investment climate statement." U.S. Department of State, July 22, 2003.

———. "Turkmenistan: Bold new production strategy." *Petroleum Economist*, Vol. 70, Iss. 11, November 2003: 32 (2).

———. "Turkmenistan faces challenges in export transportation options." *Oil & Gas Journal*, Vol. 100, Iss. 44, October 28, 2002: 48.

———. "Turkmenistan: How healthy is Niyazov?" *Eurasianet*, June 21, 2005.

———. Turkmenistan lists ambitious plans for gas, oil development." *Oil & Gas Journal*, Vol. 100. Iss. 42, October 14, 2002: 43 (4).

———. "Turkmenistan: President says society not yet ready for democracy." *FBIS*, October 26, 2003.

———. "Ukrainian, Georgian premiers agree on Odessa-Brody oil pipeline usage." *FBIS*, February 28, 2005.

———. "UPDATE: Bulgaria, Greece Russia sign deal on Balkan pipe." *Dow Jones Newswires*, April 12, 2005.

———. "Uzbek government forces closure of local SOROS foundation." *Press Release—Open Society Institute*, April 18, 2004.

———. "Uzbekistan hopes to be key gas exporter." *Interfax*, October 2004.

———. "Uzbekistan: No good political options in sight." *Eurasianet*, May 18, 2005.

———. "West-East Gas Transmission Project." *Infopetro.com*, 2003.

———. "Work on Nabucco pipeline gets underway." *Alexander's Gas & Oil Connections*, Vol. 10, Iss. 16, August 31, 2005.

———. "Worldwide look at reserves and production." *Oil & Gas Journal*, Vol. 103, Iss. 47, December 19, 2005: 24 (2).
———. "Worldwide look at reserves and production." *Oil & Gas Journal*, Vol. 102, Iss. 47, December 20, 2004: 22 (2).
———. "Worldwide look at reserves and production." *Oil & Gas Journal*, Vol. 101, Iss. 49, December 22, 2003: 46 (2).
———. "Worldwide reserves." *Oil & Gas Journal*, Vol. 100, Iss. 52, December 23, 2002: 114 (2).
———. "Worldwide reserves." *Oil & Gas Journal*, Vol. 94, Iss. 53, December 30, 1996: 40 (2).
———. "Year in review 2003: World-Affairs Georgia." *Encyclopedia Britannica Online*, 2004.
———. "Year in review 2003: Uzbekistan." *Encyclopedia Britannica Online*, 2004.
Antelava, Natalia, and Dima bit-Suleiman. "Oil is not enough?" *Transitions Online*, September 25, 2003.
——— and Giorgi Lomsadze. "Radical opposition party tops public opinion poll in Georgia as parliament election looms." *Eurasianet*, October 28, 2003.
Areshidze, Irakly. "Questions for Mikheil Saakashvili." *The Washington Post*, February 21, 2004: A19.
Arvedlund, Erin. "Russia weighs tax shift toward oil sector." *The New York Times*, April 9, 2004: W1.
Asian Development Bank. *Technical Assistance for the Feasibility Studies of the Turkmenistan-Afghanistan-Pakistan Natural Gas Pipeline Project.* December 2002: 1 (15).
Askari, Hossein. "Caspian oil development: The sooner the better." *Business Economics*, Vol. 38, Iss. 2, April 2003: 43 (7).
Bahgat, Gawdat. *American Oil Diplomacy in the Persian Gulf and the Caspian Sea.* Gainesville, FL: University of Florida Press, 2003.
Baker, Peter. "Called home to rebuild, Georgian led revolution; Reforms expected of U.S.-educated Saakashvili." *The Washington Post*, November 26, 2003: A1.
——— and Susan B. Glasser. "Push for democracy has authoritarians unnerved." *The Washington Post*, January 26, 2004: A11.
Baker and McKenzie. "CIS legislative service." May/June 2003: 10.
Bakoulev, Pavel. "PSA: Still alive in Russia?" *Caspian Investor*, Vol. 7, Iss. 4, April 2004: 27.
Bamber, Derek. "Venturing into a new world." *Petroleum Economist*, Vol. 70, Iss. 12, December 2003: 11 (2).
Barnes, Joe. "U.S. national interests: Getting beyond the hype." In Yelena Kalyuzhnova, Amy Myers Jaffe, Dov Lynch, and Robin C. Sickles, eds., *Energy in the Caspian Region: Present and Future*, 212–233. Hampshire, UK, and New York: Palgrave Publishers Ltd, 2002.
———. and Ronald Soligo. "Baku-Ceyhan pipeline: Bad economics, bad politics, bad idea." *Oil & Gas Journal*, Vol. 96, Iss. 43, October 26, 1998: 29 (5).
Bayatly, Tamam. "BP current developments: First pipe for BTC pipeline arrives in Baku." *Azerbaijan International*, Spring 2003.
Berman, Ilan. "The great game's newest player speaks Chinese." *The Wall Street Journal Online*, November 14, 2003.

Berniker, Mark. "In Georgia's 'Rose Revolution,' red ink remains." *Eurasianet*, December 12, 2003.

Bielecki, J. "Energy security: Is the wolf at the door?" Special Issue: Oil and the Economy: Recent Developments in Historical Perspective. *The Quarterly Review of Economics and Finance*, Vol. 42. Iss. 2, 2002: 235 (16).

Billmeier, Andreas, Jonathan Dunn, and Bert van Selm. *In the Pipeline: Georgia's Oil and Gas Transit Revenues*. IMF Working Paper WP/04/209. November 2004: 8 (16).

Blagov, Sergei. "Delay, competition mar Russia's Turkmenistan venture." *Eurasianet*, December 19, 2003.

———. "Russia and Armenia: United by geopolitics, divided by energy resources." *Eurasianet*, May 17, 2004.

———. "Russia outmaneuvers Ukraine for Turkmen gas—For now." *Eurasianet*, April 21, 2005.

———. "Russian oil companies to raise oil output as OPEC plans cut." *OPECNA News Service*, February 12, 2004.

Blandy, Charles. "The Caspian: A catastrophe in the making—The destruction of a unique ecosystem." Sandhurst, UK: Conflict Studies Research Centre, 1997.

Blua, Antoine. "Caspian: Influx of killer jellyfish threatens fish stocks." *Eurasianet*, July 3, 2004.

———. "Uzbekistan: EBRD to decide on future relationship." *RFE/RL*, April 5, 2004.

Blum, Douglas. "The Russian trade-off: Environment and development in the Caspian Sea." *Journal of Environment & Development*, Vol. 7, Iss. 3, September 1998: 248 (30).

BP. "Azeri, Chirag & Gunashli Full Field Development Phase I." Azerbaijan Business Unit. *Executive Summary. Azeri, Chirag & Gunashli Full Field Development Phase I*. February 2002.

———. "Caspian development and export bulletin: Regional review." Section 3, 4. January 2002.

———. *Statistical Review of World Energy June 2005*. London, UK: BP, 2005.

———. *Statistical Review of World Energy June 2004*. London, UK: BP, 2004.

———. *Statistical Review of World Energy June 2003*. London, UK: BP, 2003.

———. *Statistical Review of World Energy June 2002*. London, UK: BP, 2002.

BP Azerbaijan. *Sustainability Report 2004*.

Brady, Rose. "Central Asia: All in the family." *Business Week*, Iss. 3855, October 27, 2003: 63.

Brennan, Dan. "Like father, like son? Azerbaijan's bleak legacy." *Eurasianet*, December 23, 2003.

Brexendorff, Alexander, and Christian Ule. "Changes bring new attention to Iranian buyback contracts." *Oil & Gas Journal*, Vol. 102, Iss. 41, November 1, 2004: 16 (4).

Brodman, John R. "Energy (in)security in the 21st century." *IAEE Newsletter*, First Quarter 2004: 14 (5).

Brower, Derek. "Seeking export outlets." *Petroleum Economist*, Vol. 70, Iss. 11, November 2003: 34.

Brown, Frank. "Fear factor: In Vladimir Putin's Russia, worries about what comes next are stifling business." *Newsweek (International Edition)*, October 25, 2004: 28.

Browne, Andrew, Matt Pottinger, and Peter Wonacott. "China's expansion may be easing." *The Wall Street Journal*, June 11, 2004: A2.

Browne, Lord John. *Energy—the Medium-Term Outlook*. Washington, DC: Washington Press Club, March 23, 2004.

Caspian Events Limited and Azerbaijan Chamber of Commerce & Industry. "Caspian Oil & Gas 2004." Kingston-Upon-Thames, UK: 2003, 4.

Cavanna, Pietro. "Kazakhstan project." ENI Slide Presentation. October 3, 2003.

Center for Strategic & International Studies. "Implications of China's energy search." *Conference*. Washington, DC. March 23, 2005.

Central Intelligence Agency. "Azerbaijan." *2005 World Factbook*.

Chalabi, Fadhil J. "Gulf oil vs. the oil of the Caspian Sea." In *Caspian Energy Resources: Implications for the Arab Gulf*, 155–167. The Emirates Center for Strategic Studies and Research. London and New York: I. B. Tauris & Co Ltd., 2000.

Chazan, Guy. "Kremlin blocks Siemens purchase." *The Wall Street Journal*, April 14, 2005: A14.

———. "Russian taxman irks investors." *The Wall Street Journal*, November 10, 2004: A15.

ChevronTexaco. "Kazakhstan fact sheet." ChevronTexaco Website.

Chew, Ken. "Global oil supply issues: Recent trends and future possibilities." HIS Energy Group. Presentation at the APPEX London Global Perspectives Forum. March 2, 2005.

Chow, Edward C. "Russian pipelines: Back to the future?" *Georgetown Journal of International Affairs*, Vol. 5, Iss. 1, Winter/Spring 2004: 27 (7).

Christoffersen, Gaye. "China's intentions for Russian and Central Asian oil and gas." *NBR Analysis*, Vol. 9, No. 2. Seattle, WA: The National Bureau of Asian Research, March 1998.

Clark, Judy. "Gazprom, partners to rehabilitate CAC pipeline, study awarded." *Oil & Gas Journal Online*, November 12, 2003.

Clark, Martin. "No going back." *Petroleum Economist*, Vol. 70, Iss. 12, December 2003: 7 (3).

Coburn, Leonard L. Oral testimony before the U.S. Senate Foreign Relations Committee, "United States Energy Security: Russia and the Caspian." April 30, 2003.

Cole, Bernard D. *'Oil for the Lamps of China'—Beijing's 21st-Century Search for Energy*. Washington, DC: Institute for National Strategic Studies, 2003.

Council on Foreign Relations. *Iran: Time for a New Approach*. New York: Council on Foreign Relations, July 2004: 1 (100).

Cowan, Janice. "Clearing the air Kazakhstan style." *Euromoney*, Iss. 381, January 2001: 86.

Coxe, Donald. "The China syndrome." *Maclean's*, Vol. 117, Iss. 20, May 17, 2004: 29.

Crawshaw, Steve. "Clare Short helps out a tyrant." *New Statesman*, Vol. 16. Iss. 745, February 24, 2003: 17.

Croissant, Cynthia. *Azerbaijan, Oil and Geopolitics*. Commack, NY: Nova Science Publishers, Inc., 1998.

Croissant, Michael P. and Bülent Aras, editors. *Oil and Geopolitics in the Caspian Sea Region*. Westport, CT: Praeger Publishers, 1999.

Dekmejian, R. Hrair, and Hovann H. Simonian. *Troubled Waters: The Geopolitics of the Caspian Region*. London and New York: I.B. Tauris & Co Ltd, 2003.

Denison, Irina, and Ivan Grigoryev. "Pipeline's price for progress." *Caspian Investor*, Vol. 7, Iss. 5, May 2004: 18–20.

Din, Shams-Ud, editor. *Geopolitics and Energy Resources in Central Asia and the Caspian Sea Region*. Delhi, India: Lancers Books, 2000.

Ebel, Robert E. "Caspian oil: Separating myth and reality." In *Central Asia and the Caucasus: A New Version of the Great Game. The Findings of a Conference Organized by the DACOR Bacon House Foundation*, 69–79. Washington, DC: DACOR Bacon House Foundation, 1998.

———. *Energy Choices in the Near Abroad: The Haves and Have-Nots Face the Future*. Washington, DC: The Center for Strategic and International Studies, 1997.

Email correspondence. Anderson, Elizabeth. Analyst. *The Economist Intelligence Unit*. New York. February 1, 2005–February 3, 2005.

———. Askari, Dr. Hossein. Author. Iran Professor of International Business and Professor of International Affairs. George Washington University. Washington, DC. April 1, 2004.

———. Baev, Dr. Pavel. Author. International Peace Research Institute. Oslo, Norway. December 17, 2003–December 19, 2004.

———. Bahgat, Dr. Gawdat. Author, and Professor of Political Science, Indiana University of Pennsylvania. Indiana, Pennsylvania. October 6, 2003–October 7, 2003.

———. Bakhtiari, Dr. A. M. Samsam. Author. Senior Expert at the National Iranian Oil Company (NIOC). Tehran, Iran. October 22, 2003–January 16, 2004.

———. Billingsley, Tara. Analyst. U.S. Department of Energy. Washington, DC. March 11, 2004.

———. Broad, Oliver. BP Webmaster. October 27, 2003–November 23, 2005.

———. CIA Analyst #1. October 10, 2003–December 10, 2004.

———. CIA Analyst #2. March 9, 2004–December 13, 2004.

———. CIA Analyst #3. January 18, 2004.

———. Coburn, Leonard. Author. U.S. Department of Energy. Washington, DC. March 11, 2004–December 6, 2005.

———. Cohen, Michael. Analyst. DOE/EIA. Washington, DC. June 30, 2005–February 24, 2006.

———. Correll, David. Analyst. DOE/EIA. Washington, DC. August 29, 2003–April 12, 2004.

———. Creswell, Laura. Analyst. Organization for Economic Cooperation and Development. Washington, DC, office. April 14, 2004–November 29, 2005.

———. DuPuis, Jean-Pierre. Analyst. Organization for Economic Cooperation and Development. International Energy Agency. Paris, France. March 5, 2004.

———. Engur, Emre. Author. BOTAS. Istanbul, Turkey. December 24, 2003.

———. Epstein, Richard. Analyst. The Economist Intelligence Unit. New York. February 1, 2005.
———. Esser, Charles. Analyst. DOE/EIA. Washington, DC. June 21, 2004.
———. Feld, Lowell. Analyst. DOE/EIA. Washington, DC. June 18, 2004–June 22, 2004, and February 21–24, 2006.
———. Finley, Mark. BP. Washington, DC. October 26, 2004–December 13, 2004.
———. Former U.S. Ambassador to Kazakhstan. Washington, DC. September 3, 2003.
———. Forsythe, Rosemarie. Author. Senior Government Relations and Issues Advisor. ExxonMobil Corporation. Irving, TX. September 3, 2003–September 4, 2003.
———. Girodet, Catherine. Analyst. Organization for Economic Cooperation and Development. International Energy Agency. Paris, France. March 5, 2004.
———. Griffin, Nick. Author. Deutsche Bank. London. January 15, 2004.
———. Hill, Joshua. Analyst. The Economist Intelligence Unit. New York. April 15, 2004.
———. Hueper, Paul. Author. International Trade Administration, U.S. Department of Commerce. June 18, 2004–June 22, 2004, and May 1–June 30, 2005.
———. International oil company representative. February 22, 2004–December 23, 2004.
———. International oil company representative. Washington, DC. October 7, 2003.
———. Johnson, Dr. Karen. Economist. Federal Reserve Board. Washington, DC. April 10, 2004–April 29, 2004.
———. Karayianni, Dr. Marika S. Author. Attorney and Black Sea Economic Conference energy expert. Istanbul, Turkey. January 15, 2004–January 16, 2004.
———. Kinman, Michelle. Extractive industry analyst. Media Coordinator. *Crude Accountability*. Los Angeles, CA. May 27, 2005–December 12, 2005.
———. Kreil, Erik. Analyst. DOE/EIA. Washington, DC. September 9, 2003–March 23, 2004.
———. Kvitsinskaia, Elena. Analyst. UNEP Regional Office for Europe. December 13, 2005, and February 27–28, 2006.
———. Lang, Fréderique. Analyst. Organization for Economic Cooperation and Development. International Energy Agency. Paris, France. March 5, 2004.
———. McCafferty, Mark. Author. Wood Mackenzie. Edinburgh, UK. February 24, 2004–March 18, 2004.
———. McCutcheon, Hilary. Author. Wood Mackenzie. Edinburgh, UK. February 24, 2004.
———. Melby, Dr. Eric D. K. Principal. The Scowcroft Group. Washington, DC. January 3, 2004–January 20, 2004.
———. Moldabekova, Aigul. Second Secretary. Embassy of Kazakhstan. Washington, DC. November 25, 2003.
———. Nanay, Julia. Author. Senior Director, Regional Services, Corporate Advisory Group. PFC Energy. Washington, DC. September 24, 2003–September 29, 2003.
———. Nichol, Dr. Jim. Author. Congressional Research Service. Library of Congress. Washington, DC. August 7, 2003–May 19, 2004.

———. Palley, Dr. Thomas. Author. Chief Economist, US–China Economic & Security Review Commission. Washington, DC. April 6, 2004–June 27, 2004.

———. Quercioli, Riccardo. Analyst. Organization for Economic Cooperation and Development. International Energy Agency. Paris, France. March 5, 2004–March 9, 2004.

———. Ramsay, Ambassador William C. Deputy Executive Director. Organization for Economic Cooperation and Development. International Energy Agency. Paris, France. January 1, 2003–February 28, 2005.

———. Rasizade, Dr. Alec. Author. Senior associate. The Historical Research Center. Washington, DC. October 22, 2003–May 31, 2005.

———. Rossi, Vanessa. Author. Oxford Economic Forum. Oxford, UK. May 18, 2004–May 19, 2004, and February 28–March 3, 2006.

———. Sabonis-Helf, Dr. Theresa. Professor of National Security Strategy. National War College. National Defense University. Washington, DC. August 28, 2003.

———. Sieminski, Adam. Author. Senior Analyst. Deutsche Bank. London. July 2003–February 2005.

———. Steiger, Julie. Analyst. Center for Eurasian, Russian and East European Studies. Georgetown University. Washington, DC. February 24, 2004.

———. Ter-Sarkarian, Dafne. Analyst. The Economist Intelligence Unit. New York. February 1, 2005–February 3, 2005.

———. U.S. government representative in Moscow. April 13, 2004–February 22, 2005.

———. Van Agt, Christof. Analyst. Office of Non-Member Countries. Organization for Economic Cooperation and Development. International Energy Agency. Paris, France. February 1, 2003–July 31, 2003.

———. Verrastro, Frank A. Director. CSIS Energy Program. Center for Strategic and International Studies. Washington, DC. April 12, 2004.

———. Wasylkiw, Ihor P. Vice President Investor Relations. PetroKazakhstan. Calgary, Alberta, Canada. January 15, 2004–January 25, 2004.

———. Watson, Paul. Office of the Secretary of Defense. February 26, 2004–June 21, 2004.

———. Wolf, Ambassador John S. U.S. Department of State. Assistant Secretary for Nonproliferation Affairs. October 3, 2003–October 16, 2003.

———. Yang, Yongzheng. Economist. International Monetary Fund. April 10, 2004–April 13, 2004.

Emerson, Sarah A. "The relevance of Caspian oil for the world market." In *Caspian Energy Resources: Implications for the Arab Gulf*, 169–186. The Emirates Center for Strategic Studies and Research. London and New York: I. B. Tauris & Co Ltd., 2000.

Farzanegan, M. R. "Iranian options most economically viable for exporting Caspian oil." *Oil & Gas Journal*, Vol. 101, Iss. 11, March 17, 2003: 22 (4).

Fesharaki, Fereidun, and Hassaan Vahidy. "Pakistan's gas discoveries eliminate import need." *Oil & Gas Journal*, Vol. 101, Iss. 6, January 28, 2002: 32 (6).

Figure. "Decline of Caspian Sea sturgeon catch and rising waters of the Sea." United Nations Development Program. Caspian Environment Programme: UNEP/GRID Arendal, Norway.

———. "Kashagan oil field proposed production profile." Contained in Nick Griffin and Caroline Cook. *Kashagan—Catch of the Day?* Deutsche Bank. February 24, 2003: 1 (5). Private Report.

———. "Uzbekistan's natural gas supply/demand balance to 2020 (mmcf/d)." Provided by Wood Mackenzie. *Uzbekistan Country Overview*. Edinburgh, UK: Wood Mackenzie, 2003. Private Report.

———. "Uzbekistan's oil supply/demand balance to 2020 ('000 b/d)." Wood Mackenzie. *Uzbekistan Country Overview*. Edinburgh, UK: Wood Mackenzie, 2003. Private Report.

Fletcher, Sam. "US pressure disrupts Iran's S. Azadegan development." *Oil & Gas Journal*, Vol. 102, Iss. 35, September 20, 2004: 24 (2).

———. "World oil production to rise by 16 million b/d by 2010." *Oil & Gas Journal*, Vol. 103, Iss. 7, February 21, 2005: 32 (2).

Forsythe, Rosemarie. *The Politics of Oil in the Caucasus and Central Asia*. Adelphi Paper 300. International Institute for Strategic Studies. London: Oxford University Press, 1996.

Foss, Michelle Michot, S. Gürcan Gülen, and Bhamy Shenoy. "Caspian pipeline prospects hinge on transparent, logical regulatory frameworks." *Oil & Gas Journal*, Vol. 98, Iss. 34, August 21, 2000: 68 (5).

Freizer, Sabine. "Georgia's constitutional amendments: A setback for democratization?" *Central Asia–Caucasus Analyst*, February 11, 2004.

Gavin, James. "Tehran drives for transit status." *Petroleum Economist*, Vol. 70, Iss. 12, December 2003: 5 (2).

Geoffrey, Cline & Associates. "Study of the oil and gas sectors in the Caspian with special emphasis on Kazakhstan and Azerbaijan." Glasgow, UK: Trade Partners UK, April 2003.

Gladney, Dru C. "China's interest in Central Asia: Energy and ethnic security." In Robert Ebel and Rajan Menon, eds., *Energy and Conflict in Central Asia and the Caucasus*, 209–223. Lanham, MD: Rowman & Littlefield Publishers, Inc., 2000.

Glasser, Susan B. "President's son sweeps election in Azerbaijan." *The Washington Post*. October 16, 2003: A14.

———. "Rise of leader's son sharpens Azerbaijan's identity crisis." *The Washington Post*, August 9, 2003: A8.

Gokay, Bülent, editor. *The Politics of Caspian Oil*. Houndmills, UK: Palgrave Publishers Ltd, 2001.

Goldstein, Morris, and Nicholas R. Lardy. "What kind of landing for the Chinese economy?" *Policy Brief PB04-7*. November 2004. Washington, DC: Institute for International Economics.

Goldwyn, David L. "Extracting transparency." *Georgetown Journal of International Affairs*, Vol. 5, Iss. 1, Winter/Spring 2004: 5 (12).

Goodarzy, Hormoz. "Organizational response to the Caspian's environmental needs." In Hooshang Amirahdi, ed., *The Caspian Region at a Crossroad*, 137–144. New York: St. Martin's Press, 2000.

Gorst, Isabel. "Kazakhstan: Solving the export puzzle." *Petroleum Economist*, Vol. 69, Iss. 11, November 2002: 26 (3).

———. "Moscow flexes its muscles." *Petroleum Economist*, Vol. 71, Iss. 1, January 2004: 29 (2).

———. "Whips and honey cakes." *Petroleum Economist*, Vol. 71, Iss. 1, January 2004: 26 (2).

Gregory, Paul. "Developing Caspian energy reserves: The legal environment." In *Caspian Energy Resources: Implications for the Arab Gulf*, 25–52. The Emirates Center for Strategic Studies and Research. London and New York: I. B. Tauris & Co Ltd., 2000.

Griffin, Nick, and Caroline Cook. *Kashagan—Catch of the Day?* Deutsche Bank. February 25, 2004: 1 (5). Private Report.

———. *Kashagan: The Big Sturgeon*. Deutsche Bank. October 10, 2003: 1 (21). Private Report.

Grigoryev, Ivan. "Conflict of Caspian Pipeline Consortium shareholders' interests is ruining their business: Pipe ruptured by squabbling." *Caspian Investor*, Vol. 7, Iss. 2, February 2004: 21 (3).

——— and Maria Yaklovleva. "Key E&P projects in northern Caspian Sea ratchet back the speed of development: Shifting into lower gear." *Caspian Investor*, Vol. 7, Iss. 5, May 5, 2004: 27 (4).

Gu, Mengdi. "China wants more pipelines for improved oil import security." *Oil & Gas Journal*, Vol. 103, Iss. 1, January 3, 2005, 59.

Hadjian, Avedis Bedros. "Azerbiajan's energy policy and its implications for Russian security." In Bülent Gokay, ed., *The Politics of Caspian Oil*. Houndsmills, UK, and New York, NY: Palgrave, 2001.

Hansen, Sander. *Pipeline Politics: The Struggle for Control of the Eurasian Energy Resources*. The Hague, NL: The Clingendael Institute, 2003.

Hardin, Katherine. "Elections in Azerbaijan: Another Aliev takes the helm." *CERA Decision Brief*. Cambridge, MA: Cambridge Energy Resource Associates. December 2, 2003: 1 (5). Private Report.

——— and Nurzhan Kapparov. "Kazakhstan's gas dilemma." *CERA Decision Brief*. Cambridge, MA: Cambridge Energy Resource Associates. January 2004: 1 (11). Private Report.

——— and Scott Roberts. "The fourth wave: Renewed Asian interest in the Caspian." *CERA Decision Brief*. Cambridge, MA: Cambridge Energy Resource Associates. August, 2005: 1 (15). Private Report.

Hayes, Chalain. "Uzbekistan primer." USCINCPAC Virtual Information Center (VIC). September 16, 2001: 1 (79).

Hekimoglu, Levent. "Caspian oil and the environment: Curse or cure?" In Michael P. Croissant and Bülent Aras, eds., *Oil and Geopolitics in the Caspian Sea Region*, 83–98. Westport, CT: Praeger Publishers, 1999.

Hill, Fiona. "Pipeline dreams in the Caucasus." *The Caucasus and the Caspian: 1996–1997 Seminar Series*. Cambridge, MA: Harvard University, John F. Kennedy School of Government, September 1996.

———. "Uzbek fury." *The Wall Street Journal Online*, May 20, 2005.

Hiro, Dilip. "Why is the US inflating Caspian oil reserves?" *Middle East International*, September 12, 1997: 18 (2).

Hoyos, Carola, and Richard McGregor. "PetroChina ends talks on pipeline venture with foreign companies." *Financial Times*. August 4, 2004: 8.

Hurd, David. "China oil pipeline plans." *Oil & Gas Abacus: Can OPEC Pull It Off?* Deutsche Bank. November 3, 2003: 5 (27). Private Report.

Interfax Petroleum Report. June 10–17, 1994. Cited in Ebel, Robert E. *Energy Choices in the Near Abroad: The Haves and Have-nots Face the Future*. Washington, DC: The Center for Strategic and International Studies, 1997.

International Energy Agency in Co-operation with the Energy Charter Secretariat. *Caspian Oil and Gas: The Supply Potential of Central Asia and Transcaucasia*. Paris, France: OECD/IEA, 1998.

International Energy Agency. *World Energy Investment Outlook, 2003, Insights*. Paris, France: OECD/IEA, 2003: 154.

———. *World Energy Outlook 2004*. Paris, France: OECD/IEA, 2004: 82.

Interview, personal. Analysts at Central Intelligence Agency. Langley, VA. October 21, 2003.

———. Analysts at Defense Intelligence Agency. Washington, DC. September 9, 2003.

———. BP representative. Tehran, Iran. May 25, 2004.

———. Coburn, Leonard. Author. U.S. Department of Energy. Washington, DC. September 9, 2003, and February 18, 2004.

———. Former Ambassador to Kazakhstan. Washington, DC. August 29, 2003.

———. Khajehpour, Bijan. Managing Director, Atieh Bahar Consulting. Tehran, Iran. Washington, DC. March 5, 2004.

———. Mann, Ambassador Steven. Washington, DC. August 27, 2002.

———. Moldabekova, Aigul. Second Secretary. Embassy of Kazakhstan. Washington, DC. November 21, 2003.

———. Ramsay, Ambassador William C. Deputy Executive Director. International Energy Agency. Paris, France. June 20, 2003.

———. Rasizade, Dr. Alec. Author. Senior Associate. The Historical Research Center. Washington, DC. October 8, 2003.

———. Sabonis-Helf, Dr. Theresa. Professor of National Security Strategy. National War College. National Defense University. Washington, DC. August 19, 2003.

———. Saidkhodjaeva, Dr. Malika. The World Bank. Country Office in Uzbekistan. Washington, DC. July 9, 2004.

———. Sedney, Diana. ChevronTexaco. Washington, DC. October 24, 2003.

———. Stevens, Dr. Paul. Professor of Petroleum Policy & Economics. Centre for Energy, Petroleum and Mineral Law and Policy. University of Dundee. Dundee, Scotland, UK. May 22, 2004.

———. Yaphe, Dr. Judith. Senior Fellow. Institute for National Strategic Studies. National Defense University. Washington, DC. August 8, 2003.

Interview, telephone. Clawson, Dr. Patrick. Author. Deputy Director. The Washington Institute for Near East Policy. Washington, DC. September 30, 2003.

———. Collier, Charles. Analyst. U.S. Department of Energy. Washington, DC. August 12, 2003, and February 13, 2004.

———. Forsythe, Rosemarie. Author. Senior Government Relations and Issues Advisor. ExxonMobil Corporation. Irving, TX. September 3, 2003.

———. Nichol, Dr. Jim. Author. Congressional Research Service. Library of Congress. Washington, DC. August 8, 2003.

———. Phelps, Frances. Retired Unocal employee. Washington, DC. November 25, 2003.

———. Schlingemann, Frits. Director and Regional Representative. UNEP Regional Office for Europe. June 3, 2005.

———. Watson, Paul. Office of the Secretary of Defense. Washington, DC. February 18, 2004.

Iskyan, Kim. "Playboy of the Eastern World: Why you should care about Azerbaijan's presidential elections." *Slate*, October 13, 2003.

Ismailova, Gulnara. "Over-fishing of sturgeon could harm Azerbaijan's caviar export." *Central Asia–Caucasus Analyst*, June 2, 2004.

Ismailzade, Fariz. "Economic ties bind Azerbaijan and Georgia." *Eurasianet*, November 12, 2003.

Jack, Andrew. "Nazarbayev considers oil pipeline to Gulf via Iran." *Financial Times (Asia Edition)*, May 26, 2004: 2.

Jaffe, Amy Myers. "Price vs. market share for the Arab Gulf oil producers: Do Caspian oil reserves tilt the balance?" In *Caspian Energy Resources: Implications for the Arab Gulf*, 143–153. The Emirates Center for Strategic Studies and Research. London and New York: I. B. Tauris & Co Ltd., 2000.

———. "US Policy Towards the Caspian Region: Can the Wish-List Be Realized?" In Gennady Chufrin, ed., *The Security of the Caspian Sea Region*, 136–150. Oxford, UK: Stockholm International Peace Research Institute and Oxford University Press, 2001.

Kadrzhanova, Azhar. "Kazakhstan: Caspian Oil Development Plan." U.S. Department of Commerce. Commercial Service. July 3, 2003.

Kahn, Gabriel, Zaure Kistanova, and Geoffrey T. Smith. "Kazakh Government to Buy Into Huge Oil Project." *The Wall Street Journal*. January 31, 2005: A16.

Kalyuzhnova, Yelena. "Economics and Energy." In Yelena Kalyuzhnova, Amy Myers Jaffe, Dov Lynch, and Robin C. Sickles, eds., *Energy in the Caspian Region: Present and Future*, 58–85. Hampshire, UK, and New York: Palgrave Publishers Ltd., 2002.

Karagiannis, Emmanuel. *Energy and Security in the Caucasus*. London: Routledge-Curzon, 2002.

Kashfi, Mansour. "Lack of data casts doubt on Iranian reserves hike." *Oil & Gas Journal*. Vol. 102, Iss. 41. November 1, 2004: 20 (2).

Katz, Mark N. "Assessing the Saudi-Russian Summit: Much Ado About Nothing?" *Central Asia-Caucasus Analyst*. November 5, 2003.

Kemp, Geoffrey. "Iran and Caspian Energy: Prospects for Cooperation and Conflict." In *Caspian Energy Resources: Implications for the Arab Gulf*, 53–70. The Emirates Center for Strategic Studies and Research. London and New York: I.B. Tauris & Co Ltd., 2000.

Khartukov, Eugene, and Ellen Starostina. "Russia's New Pipelines Will Debottleneck Exports, Production." *Oil & Gas Journal*. Vol. 101, Iss. 38. October 6, 2003: 62 (10).

King, Charles. "A Rose Among Thorns: Georgia Makes Good." *Foreign Affairs*. Vol. 83, Iss. 2. March/April 2004: 13 (6).

———. "Potemkin Democracy: Four Myths about Post-Soviet Georgia." *The National Interest*. Iss. 64. Summer 2001: 93 (12).

Kohl, Wilfrid L. "The Development of Caspian Sea Oil: Implications for OPEC." In *Caspian Energy Resources: Implications for the Arab Gulf*, 127–142. The Emirates Center for Strategic Studies and Research. London and New York: I. B. Tauris & Co Ltd., 2000.

Kramer, Andrew E. "Moscow Moves to Restore A Cap on Price of Gasoline." *The New York Times*. February 16, 2006: C5.

Krastev, Nikola. "U.S./Uzbekistan: U.S. Figure Linked to Kazakh Leaders Faces Corruption Hearing." *RFE/RL*. June 3, 2004.

Kusainov, Aldar. "Wheels Set in Motion for Dynastic Transition in Kazakhstan." *Eurasianet*. September 16, 2003.

Laughland, John. "Georgia on Their Mind." *The Guardian*. April 1, 2004: 22.

Lee, Julian. "FSU Pipeline Advisory." *Centre for Global Energy Studies*. No. 5. March 12, 2004.

Levine, Steve. "Caspian Intrigue: Odd Family Drama in Kazakhstan Dims Democratic Hopes—Petroleum-Rich Land's Ruler Smacks Down Son-in-Law and Sundry Challengers—Oil Companies Feel Heat, Too." *The Wall Street Journal*. September 12, 2002: A1.

———. "Kazakhstan Hangs Tough in Oil Flap—Clash With ChevronTexaco and ExxonMobil Reveals New Approach to Foreigners." *The Wall Street Journal*. November 22, 2002: A11.

———. "Turkmenistan's Allegations May Roil Central Asia." *The Wall Street Journal*. December 26, 2002: A8.

Lichtblau, John. "U.S. Policy and Action Toward Central Asia and the Caspian Sea Countries." In Dieter Dettke, ed., *A Great Game No More: Oil, Gas and Stability in the Caspian Sea Region*, 67–71. Washington, DC, and Berlin, Germany: Friedrich-Ebert-Stifting, 1999.

Lipman, Marsha. "Birth of a Dynasty?" *The Washington Post*. August 11, 2003: A17.

Lobe, Jim. "Uzbek Leader Becoming Embarrassment to West." *OneWorld.net*. December 8, 2003.

Love, Patrick. "The Changing Face of Energy Politics." *The OECD Observer*. No. 217/218. Summer 1999.

Lubin, Nancy. "Turkmenistan's Energy: A Source of Wealth or Instability?" In Robert Ebel and Rajan Menon, eds., *Energy and Conflict in Central Asia and the Caucasus*, 107–121. Oxford, UK and Lanham, MD: Roman & Littlefield Publishers, Inc., 2000.

MacWilliam, Ian. "Kazakh-China Oil Pipeline Opens." *BBC News*. December 15, 2005.

Mann, Steven R. "Caspian Futures." In Jan H. Kalicki and Eugene K. Lawson, eds., *Russian-Eurasian Renaissance*? 147–160. Washington, DC, and Stanford, CA: Woodrow Wilson Center Press and Stanford Business Books, 2003.

Map. "Caspian Region Oil and Gas Pipeline Export Routes." International Energy Agency. *World Energy Investment Outlook 2003*. Paris, France: Organization for Economic Cooperation and Development, 2003: 154.

———. "Chinese Market Possibilities for Central Asian and Russian Gas." David A. Pietz. "Special Report: China's Emerging Gas Industry." *Oil & Gas Journal*. Vol. 98, Iss. 23. June 5, 2000: 60 (7).

———. "Natural Gas Export Pipelines of Turkmenistan and Kazakhstan." International Energy Agency. *Caspian Oil and Gas: The Supply Potential of Central Asia and Transcaucasia*. Paris, France: Organization for Economic Cooperation and Development. International Energy Agency, 1998: 26.

———. "Pipelines Serving Tengiz and Karachaganak Fields." ChevronTexaco Website.

———. "Proposed Gas Pipeline from Turkmenistan through Afghanistan to Pakistan." Adapted from Asian Development Bank. *Technical Assistance for the Feasibility Studies of the Turkmenistan-Afghanistan-Pakistan Natural Gas Pipeline Project*. December 2002: 3 (15).

———. "Turkmenistan Oil and Gas Fields in or Near the Caspian Sea." Anonymous. "Foreign PSAs Help Boost Turkmen Oil Production." *Oil & Gas Journal*. Vol. 100, Iss. 43. October 21, 2002: 49 (3).

Marsh, Peter. "Long-serving Kazakh leader to stand again." *Financial Times*. July 3, 2003: 10.

McCafferty, Mark, and Valentina Kretzschmar. "CPC, BTC Pipelines Make Current Caspian Area Oil Export Capacity Adequate." *Oil & Gas Journal*. Vol. 102, Iss. 18. May 10, 2004: 18 (8).

McCutcheon, Hilary and Richard Osbon. "Risks Temper Caspian Rewards Potential." *Oil & Gas Journal*. Vol. 99, Iss. 52. December 24, 2001: 22 (5).

McElvoy, Ann. "Ashkhabad Guy: Niyazov's Nasty Reign." *The New Republic*. Vol. 210, Iss. 13. March 28, 1994: 15 (2).

McFaul, Michael, Nikolai Petrov, and Andrei Ryabov. *Between Dictatorship and Democracy: Russian Post-Communist Political Reform*. Washington, DC: Carnegie Endowment for International Peace and Brookings Institution Press, 2004.

McLaughlin, Katy. "For Caviar Lovers, It's Time to Gulp." *The Wall Street Journal*. April 21, 2004: D3.

Millennium Challenge Corporation. "Georgia and Millennium Challenge Corporation: Building a Partnership for Poverty Reduction Through Sustainable Economic Growth." September 12, 2005.

Ministry of Energy and Mineral Resources of the Republic of Kazakhstan. *Brief Extract, State Program for the Development of the Kazakhstan Sector of the Caspian Sea*. Astana, Kazakhstan: Ministry of Energy and Mineral Resources, Republic of Kazakhstan, May 2003: 1 (30).

Muller, Friedmann. "Policy Options in the Caspian Sea Region." *A Great Game No More: Oil, Gas and Stability in the Caspian Region*. Dieter Dettke, ed. Washington, DC and Berlin, Germany: Friedrich-Ebert-Stifting, 1998.

Mulvey, Stephen. "Profile: Ilham Aliyev." *BBC News Online*. October 15, 2003.

Namazi, Siamak. "The Caspian's Environmental Woes." In Hooshang Amirahdi, ed., *The Caspian Region at a Crossroad*, 121–136. New York: St. Martin's Press, 2000.

Nanay, Julia. "The North Caspian PSA: Kashagan's Development." *Middle East Economic Survey*. Vol. XLVII, No. 32. August 9, 2004.

Nash, Roland. "Taxing time for oil," *Alexander's Gas & Oil Connections*, Vol. 9 Iss. 3. February 10, 2004.

National Bank of Kazakhstan. "International Reserves and Assets of the National Oil Fund of Republic (sic) of Kazakhstan."

National Intelligence Council Conference Report. *Central Asia and the South Caucasus: Reorientations, Internal Transitions, and Strategic Dynamics*. Washington, DC: October 2000.

Neff, Andrew. "Caspian Nations Pursuing Oil Exports at Greatly Varying Paces." *Oil & Gas Journal*. Vol. 103, Iss. 22. June 13, 2005: 38.

Newberg, Paula R. "Oil-Lubricated Realpolitik Deepens Divisions in an Ailing Caucasus." *The Los Angeles Times*. October 12, 2003: M2.

OGJ editors. "Study Criticizes Corporate Governance in Russia." *Oil & Gas Journal Online*. October 5, 2004.

Olcott, Martha Brill. "In Uzbekistan, the Revolution Won't Be Pretty." *The Washington Post*. May 22, 2005: B2.

———. *Kazakhstan: Unfulfilled Promise*. Washington, DC: Carnegie Endowment for International Peace and the Brookings Institution Press, 2002.

Oster, Shai and Patrick Barta. "About Face: China Stumbles In Attempt to Cut Use of Coal and Oil." *The Wall Street Journal*. March 3, 2006: A1.

O'Sullivan, Stephen, Pavel Kushnir, and Olga Danilenko. *PetroKazakhstan—A Chinese Takeaway?* United Financial Group and Deutsche Bank. August 4, 2004: 1 (30). Private Report.

Pala, Christopher. "Investors Recoil From Oil Terms in Kazakhstan." *The New York Times*. October 27, 2004: W1.

Parker, Leia. "CNPC in $4.18 Billion Deal to Buy PetroKazakhstan." *The Wall Street Journal Online*. August 22, 2005.

Peimani, Hooman. "Central Asian States Increase Energy Swap Deals with Iran." *Central Asia-Caucasus Analyst*. June 30, 2004.

———. "Iran Intent on Keeping Pace with Caspian Basin Energy Development," *Eurasianet*. August 19, 2003.

———. "Iran Launches Two Caspian Initiatives." *Central Asia-Caucasus Analyst*. June 15, 2005.

———. *The Caspian Pipeline Dilemma: Political Games and Economic Losses*. Westport, CT: Praeger Publishers, 2001.

Perdue, Jeanne M. "Technology Credited for New Records." *Drilling and Production Yearbook*. Houston, TX: Hart Energy Publishing, LP, March 2003.

PetroKazakhstan. "Corporate Presentation CIBC World Markets Institutional Investor Conference." Whistler, BC, Canada: February 19, 2004.

———. Peters & Co. Limited 2004 North American Oil & Gas Conference. September 15, 2004. Toronto, Ontario, Canada.

Pietz, David A. "China's Emerging Gas Industry." *Oil & Gas Journal*. Vol. 98, Iss. 3. June 5, 2000: 60 (3).

Pope, Hugh. "Kazakhstan Balancing Act Gives China a Key Oil Role." *The Wall Street Journal*. November 21, 2003: A10.

Rasizade, Alec. "Azerbaijan after a Decade of Independence: Less Oil, More Graft and Poverty." *Central Asian Survey*. Vol. 21, Iss. 4. (2004): 349 (22).

———. "Georgia Meets the Test of Its Independence." *Contemporary Review*. Vol. 284, Iss. 1658. March 2004: 139 (10).

Rimple, Paul. "Azerbaijan: Taking Stock of Sturgeon." *Eurasianet*. February 26, 2006.

Romano, Cesare P. R. "The Caspian Sea Dispute: The Role of International Law." Dundee, UK: University of Dundee, Center for Energy, Petroleum and Mineral Law and Policy, January 2000.

Rossett, Claudia. "Potentate Jr." *The Wall Street Journal.* November 6, 2002: A22.

Ruseckas, Laurent. "Caspian Oil Development: An Overview." In *Caspian Energy Resources: Implications for the Arab Gulf,* 13–23. The Emirates Center for Strategic Studies and Research. London and New York: I. B. Tauris & Co Ltd., 2000.

Saidazimova, Gulnoza. "Uzbekistan: Islam Karimov vs. the Clans." *Eurasianet.* April 23, 2005.

Salameh, Mamdouh G. "Caspian Sea Is No Middle East." *International Association for Energy Economics Newsletter.* Third Quarter 2002: 12.

———. "Middle East Oil: What's the Alternative?" *Dialogue.* United States Association for Energy Economics. Vol. 11, No. 3. November 2003: 14.

Sanger, David E. "U.S. Challenge in Iran." *The New York Times.* June 26, 2005: 1:10.

Schlumberger Corporation Website.

Shenoy, Bhamy, and William James. "Caspian Energy Exports: Turkmenistan Fumbling Opportunities Afforded by Trans-Caspian Pipeline." *Oil & Gas Journal.* Vol. 98, Iss. 22. May 28, 2001: 70 (4).

Sieminski, Adam, Nicholas Aldridge, Paul Sankey, and David Hurd. *Oil and Gas Abacus: Promise of Spring.* Deutsche Bank. February 23, 2005: 1 (35). Private Report.

Skagen, Ottar. *Caspian Gas.* London, UK: Royal Institute of International Affairs, 1997.

Smith, Robert. "Politics, Production Levels to Determine Caspian Area Energy Export Options." *Oil & Gas Journal.* Vol. 99, Iss. 22. May 28, 2001: 33 (6).

SOFAZ. "SOFAZ Revenue and Expenditure Statement for 2005."

Soligo, Ronald, and Amy Myers Jaffe. "The Economics of Pipeline Routes: The Conundrum of Oil Exports from the Caspian Basin." Baker Institute Working Paper. *Unlocking the Assets: Energy and the Future of Central Asia and the Caucasus.* Houston, TX: Baker Institute, April 1998.

Somers, Andrew. "Building Blocks for Increased Trade and Investment." *American Chamber of Commerce in Russia,* February 2004.

Starobin, Paul. "Marching into the Unknown: Georgia Will Be a Test Case of the New U.S. Policy to Shore Up Strategic but Failing States." *Business Week.* Iss. 3804. October 21, 2002: 58.

Stauffer, Thomas R. "Caspian Fantasy: The Economics of Political Pipelines." *The Brown Journal of World Affairs.* Vol. VII, Iss. 263. Summer/Fall 2000: 63 (16).

Stern, David. "Uzbekistan Offers Rich Pickings for Leader's Daughter." *Financial Times.* August 19, 2003: 8.

Stevens, Paul. "Resource Impact—Curse or Blessing? A Literature Survey." *CEPMLP Internet Journal Draft Working Paper.* April 22, 2003: 1 (35).

Tavernise, Sabrina, and Christopher Pala. "Energy-Rich Kazakhstan Is Suffering Growing Pains." *The New York Times.* January 4, 2003: C1.

Telvick, Marlena, "Kazakhstan: United States vs. James H. Giffen." *Eurasianet.* May 31, 2004.

Templer, Robert. "Tashkent Dispatch: Steppe Back." *New Republic.* August 18, 2003: 11.

Teshebayeva, Dilyara. "Kazakhstan Plans Vast Expansion of Payments to Elderly, Students, and State-Sector Employees." *Eurasianet*. May 5, 2005.
Tetelashvili, Nino. "Georgia's Revolution Takes a Dangerous Turn." *Central Asia-Caucasus Analyst*. May 5, 2004.
The Commission on America's National Interests. *American National Interests*. Cambridge, MA: Harvard University, John F. Kennedy School of Government, Belfer Center for Science and International Affairs. July 2000: 1 (62).
The Economist Intelligence Unit. *Country Report on Georgia*. August 2005.
———. *Country Report on Georgia*. May 2005.
———. *Country Report on Turkmenistan*. October 2003.
———. *Country Report on Turkmenistan*. July 2003.
Timmons, Heather. "Kazakhstan Wants a Piece of Its Oil Field." *New York Times*. June 29, 2004: W1.
Torbakov, Igor. "The Geopolitical Stakes are High as Azerbaijan Holds Presidential Elections."*Eurasianet*. October 15, 2003.
Townsend, David, and Duncan Rushworth. "The Caspian Angle." *Petroleum Economist*. Vol. 68, Iss. 9. September 2001: 28.
Trade Partners UK. *Kazakhstan Oil and Gas Report*. August 2002.
Tucker, Reed. "The Five-Minute Guide to . . . the World's Remaining Tyrants." *Esquire*. Vol. 140, Iss. 2. August 2003: 110 (2).
Turkmenistan Project Open Society Institute. *Turkmenistan Weekly News Brief*. Iss. 7 (2006). February 10–16, 2006.
U.S. Department of Energy. Energy Information Administration. *Annual Energy Outlook 2005*. January 2005.
———. *Annual Energy Outlook 2004*. January 2004.
———. *Azerbaijan: Production-Sharing Agreements*. September 2004.
———. *Country Analysis Brief*. "Baltic Sea Region." January 2004.
———. *Country Analysis Brief*. "Caspian Sea Region: Environmental Issues." February 2003.
———. *Country Analysis Brief*. "Caspian Sea Region: Key Oil and Gas Statistics." August 2003.
———. *Country Analysis Brief*. "Iran." March 2005.
———. *Country Analysis Brief*. "Iran." November 2003.
———. *Country Analysis Brief*. "Kazakhstan." November 2004.
———. *Country Analysis Brief*. "Turkey." July 2005.
———. *Country Analysis Brief*. "Turkey." May 2004.
———. *International Energy Outlook 2005*.
———. *International Energy Outlook 2004*.
———. *International Energy Outlook 1998*.
———. *Monthly Energy Review*. November 2005. Table 3.3g, 54.
———. *Short Term Energy Outlook*. "OPEC Oil Production." September 2005.
U.S. Department of Energy, Office of Policy. *U.S.-Russia Energy Working Group Overview and Activities*. October 2005: 1(8).
U. S. Department of State. "Background Note: Iran." August 2004.
———. Barbara Larkin. *Caspian Region Energy Development Report, As Required By H.R. 3610*. Congressional Record. May 8, 1997. S4208.

———. *Joint Statement by President George W. Bush and President Vladimir V. Putin on U.S.-Russian Energy Cooperation*. February 24, 2005.
———. *Joint Statement by President George W. Bush and President Vladimir V. Putin on the New U.S.-Russian Energy Dialogue*. May 24, 2002.
———. Unclassified Cables: Baku 001410, July 17, 2003; Baku 001487, July 28, 2003; Almaty 005736, December 5, 2003; Almaty 000038, January 7, 2004.
U.S. Government Map. *Energy Facilities of Uzbekistan*. 734577 (R00161). March 1995.
———. *Proposed Bosporus Bypass Pipelines*. 766111AI (R02737)/392839AI. March 2004.
———. *Selected Caspian Crude Oil Export Pipelines to the Black Sea and the Mediterranean*. 764917AI (C00733). November 2003.
———. *Turkmenistan's Natural Gas Export Route to Ukraine*. (INR) 8955. January 2004.
United Nations Environment Programme. "Framework Convention for the Protection of the Marine Environment of the Caspian Sea." November 4, 2003.
———. Caspian Environment Program. UNEP/GRID-Arendal. Map and Graphic.
United Press International, "Azerbaijan Struggles to Prevent 'Dutch Disease.'" *World Peace Herald*. August 5, 2005.
United States. Executive Office of the President. The White House. *A National Security Strategy for a New Century*. Washington, DC: May 1997.
———. "Georgian Finance Minister and Millennium Challenge Corporation Vice President Sign $4.1 Million Grant." *Embassy in Georgia News*. July 17, 2005.
Van der Leeuw, Charles. *Azerbaijan: A Quest for Identity*. New York: St. Martin's Press, 2000.
———. Iran peeks beyond the fence." *Caspian Business News*. August 4, 2003.
———. "Kazakhs Unveil Oil Plan." *Caspian Business News*. June 30, 2003.
———. *Oil and Gas in the Caucasus & Caspian: A History*. New York: St. Martin's Press, 2000.
van der Schriek, Daan. "Saakashvili's First 100 Days: Bloom Still on Reforms, But Critics Voice Concern." *Eurasianet*. May 10, 2004.
Vershinin, Alexander. "Turkmenistan and China to sign gas deal. *Alexander's Gas & Oil Connections*. Vol. 10, Iss. 23. December 8, 2005.
Wallechinsky, David. "Who Is The World's Worst Dictator? A New Ranking of the Contemptible." *The Washington Post Parade Magazine*: February 13, 2005: 5.
Watkins, Eric. "Kazakhstan, Azerbaijan to transport Kazakh crude through BTC pipeline." *Oil & Gas Journal Online*. June 6, 2003.
———. "Watching the World: Spies, Taxes, and Business." *The Oil & Gas Journal*. Vol. 103, Iss. 16. April 25, 2005: 44.
Webb, John. "Russia's Offshore Caspian Projects: New Dynamics in the Wake of Demarcation Awards." *CERA Decision Brief*. Cambridge, MA: Cambridge Energy Resource Associates, August 2003: 1 (12). Private Report.
Weisbrode, Kenneth. *Central Eurasia: Prize or Quicksand?* Adelphi Paper 338. International Institute for Strategic Studies. London and New York: Oxford University Press, 2001.

White, Gregory L. "Khodorkovsky Declared Guilty and Sentenced to 9 Years in Jail." *The Wall Street Journal Online*. May 31, 2005.

———. "Moscow Restricts West's Oil Titans." *The Wall Street Journal*. February 11, 2005: A2.

Wong, James F. "China's Booming Economy Needs Energy, Pipelines, and Technology." *Oil & Gas Journal*. Vol. 101, Iss. 21. May 26, 2003: 55.

Wood Mackenzie. *Azerbaijan: Country Overview*. Edinburgh, UK: Wood Mackenzie, May 2003. 1 (84). Private Report.

———. *Kazakhstan: Country Overview*. Edinburgh, UK: Wood Mackenzie, November 2003. 1 (86). Private Report.

———. *Kazakhstan: Country Overview*. Edinburgh, UK: Wood Mackenzie, November 2002. 1 (88). Private Report.

———. *Turkmenistan: Country Overview*. Edinburgh, UK: Wood Mackenzie, May 2004. 1 (77). Private Report.

———. *Turkmenistan: Country Overview*. Edinburgh, UK: Wood Mackenzie, May 2003. 1 (82). Private Report.

———. *Uzbekistan: Country Overview*. May 2003. Edinburgh, UK: Wood Mackenzie, 2003. 1 (70). Private Report.

Zeihan, Peter. "China's Long March to Bankruptcy." *Stratfor*. March 23, 2005.

Zonn, Igor. "The Caspian Sea: Threats to Its Biological Resources and Environmental Security." In Gennady Chufrin, ed., *The Security of the Caspian Sea Region*, 69–82. Oxford, UK: Stockholm International Peace Research Institute and Oxford University Press, 2001.

Index

About the Author

MAUREEN S. CRANDALL is currently Professor of Economics at the National Defense University in Washington, D.C. She spent a number of years in the U.S. intelligence community, where she won the Director of Central Intelligence's National Intelligence Medal of Achievement. She also served on the Secretary of Energy's Nuclear Energy Research Advisory Commission, and on committees of the U.S. Association for Energy Economics. She is the author of numerous classified papers and briefings on energy, and has published several articles in defense and civilian journals.